Molecular Advancements in Tropical Diseases Drug Discovery

Molecular Advancements in Tropical Diseases Drug Discovery

Edited by

Gauri Misra, PhD

Scientist Grade-II
National Institute of Biologicals
Noida, Uttar Pradesh, India

Vijay Kumar Srivastava, PhD

Assistant Professor
Amity Institute of Biotechnology
Amity University Rajasthan
Jaipur, India

ACADEMIC PRESS
An imprint of Elsevier

ELSEVIER

Academic Press is an imprint of Elsevier
125 London Wall, London EC2Y 5AS, United Kingdom
525 B Street, Suite 1650, San Diego, CA 92101, United States
50 Hampshire Street, 5th Floor, Cambridge, MA 02139, United States
The Boulevard, Langford Lane, Kidlington, Oxford OX5 1GB, United Kingdom

Notices
Knowledge and best practice in this field are constantly changing. As new research and
experience broaden our understanding, changes in research methods, professional
practices, or medical treatment may become necessary.

Practitioners and researchers must always rely on their own experience and knowledge in
evaluating and using any information, methods, compounds, or experiments described
herein. In using such information or methods they should be mindful of their own safety
and the safety of others, including parties for whom they have a professional
responsibility.

To the fullest extent of the law, neither the Publisher nor the authors, contributors, or
editors, assume any liability for any injury and/or damage to persons or property as a
matter of products liability, negligence or otherwise, or from any use or operation of any
methods, products, instructions, or ideas contained in the material herein.

Library of Congress Cataloging-in-Publication Data
A catalog record for this book is available from the Library of Congress

British Library Cataloguing-in-Publication Data
A catalogue record for this book is available from the British Library

ISBN: 978-0-12-821202-8

For information on all Academic Press publications visit our website at
https://www.elsevier.com/books-and-journals

Publisher: Stacy Masucci
Acquisitions Editor: Rafael E. Teixeira
*Editorial Project Manager:*Mona Zahir
Production Project Manager: Kiruthika Govindaraju
Cover Designer: Mark Rogers

Typeset by TNQ Technologies

Working together
to grow libraries in
developing countries

www.elsevier.com • www.bookaid.org

Dedicated to
The lotus feet of Shri Radha Krishna
and
Goverdhannathji

Contents

Contributors

Anupam Jyoti, PhD
Amity Institute of Biotechnology, Amity University Rajasthan, Jaipur, India

Sanket Kaushik, PhD
Amity Institute of Biotechnology, Amity University Rajasthan, Jaipur, India

Gaayathri Kumarasamy, MSc
Institute for Research in Molecular Medicine (INFORMM), Universiti Sains Malaysia, Gelugor, Penang, Malaysia

Dusit Laohasinnarong, DVM, GradDip(IT), PhD, DTV, DTBVM, BM
Assistant Professor, Department of Clinical Sciences and Public Health, Faculty of Veterinary Science, Mahidol University, Salaya, Nakhon Pathom, Thailand

Yee Ling Ng
Institute for Research in Molecular Medicine (INFORMM), Universiti Sains Malaysia, Gelugor, Penang, Malaysia

Rahmah Noordin, PhD
Institute for Research in Molecular Medicine (INFORMM), Universiti Sains Malaysia, Gelugor, Penang, Malaysia

Nurulhasanah Othman, PhD
Institute for Research in Molecular Medicine (INFORMM), Universiti Sains Malaysia, Gelugor, Penang, Malaysia

Nilakshi Samaranayake, MBBS, PG Dip (Med Micro), PhD
Department of Parasitology, Faculty of Medicine, University of Colombo, Colombo, Sri Lanka

Vivornpun Sanprasert, PhD
Assistant Professor, Department of Parasitology, Faculty of Medicine, Chulalongkorn University, Pathumwan, Bangkok, Thailand; Lymphatic Filariasis and Tropical Medicine Research Unit, Chulalongkorn Medical Research Center, Faculty of Medicine, Chulalongkorn University, Pathumwan, Bangkok, Thailand

Vijay Kumar Srivastava, PhD
Amity Institute of Biotechnology, Amity University Rajasthan, Jaipur, India

Sivapong Sungpradit, PhD
Assistant Professor, Department of Pre-clinic and Applied Animal Science, Faculty of Veterinary Science, Mahidol University, Phutthamonthon, Nakhon Pathom, Thailand; The Monitoring and Surveillance Center for Zoonotic Diseases in Wildlife and Exotic Animals, Faculty of Veterinary Science, Mahidol University, Phutthamonthon, Nakhon Pathom, Thailand

Nikunj Tandel, PhD thesis
Institute of Science, Nirma University, Ahmedabad, Gujarat, India

Rajeev K. Tyagi, PhD
Ramalingaswami Fellow and Faculty, Division of Cell Biology and Immunology,
Biomedical Parasitology and Nano-immunology Lab, CSIR-Institute of Microbial
Technology (IMTECH), Chandigarh, India

Jorim Anak Ujang
Institute for Research in Molecular Medicine (INFORMM), Universiti Sains
Malaysia, Gelugor, Penang, Malaysia

Foreword

I am extremely delighted to write this Foreword for the book entitled *"Molecular Advancements in Tropical Diseases Drug Discovery"* being compiled and edited by Dr. Gauri Misra and Dr. Vijay Srivastava.

As is clear from the title, this book aims at harnessing the advances made in molecular biology of important pathogens causing amoebiasis, leishmaniasis, tuberculosis, lymphatic filariasis, malaria, and sleeping sickness for development of better diagnostics, newer drugs, and vaccines to better treat, control, and eliminate these diseases. Contributing authors are very distinguished experienced colleagues from India, Sri Lanka, Thailand, Malaysia, and the United States with original contributions to their credit. This is reflected by their insight in identification of needs, knowledge, and application.

This book has quite informative chapters on amoebiaisis, leishmaniasis, tuberculosis, lymphatic filariasis, malaria, and sleeping sickness. These diseases have been important causes of morbidity and mortality mainly in the populations of developed world. Five of these six selected tropical diseases covered in this book are parasitic diseases that continue to pose public health challenges of specific diagnosis, desired response to therapy, and lack of effective vaccines. Analysis presented in this book on current knowledge about the genetics/epigenetics of these diseases as well as causative pathogens will help the readers to think of newer strategies/targets for developing easy-to-implement, cost-effective tools for diagnosis, therapy, and vaccines for these diseases. Chapter on tuberculosis focuses on detection of infection, transmission, next-generation drugs, and their molecular action and new pathways that can be explored in near future. These topics have global interest and are covered by authors across several countries, from endemic to nonendemic countries, thus are expected to be realistic and at the same time dispassionate in analysis and opinion.

I am optimistic that this book will enlighten a wide spectrum of readers and will empower them to pursue research on aspects that they may like. I am also sure that this book will not only provide easy-to-assimilate knowledge on the biology of these diseases but also stimulate the desired interest in readers.

Finally, I compliment the authors as well as editors for this excellent effort. I wish readers an enjoyable reading, which will also help them to contribute to the development of better translational science to tackle these important diseases of immense public health importance.

(VM Katoch)

Dr VM Katoch, MD, FNASc, FASc, FAMS, FNA
NASI-ICMR Chair on Public Health Research at
Rajasthan University of Health Sciences (RUHS), Jaipur,
President, JIPMER, Puducherry,
Former Secretary, Department of Health Research, Govt of India
and Director-General, Indian Council of Medical Research,
Sector 18, Kumbha Marg, Pratap Nagar, Jaipur-302033 (Rajasthan)
E-mail: vishwamohankatoch18@gmail.com

Preface

Gauri Misra

Vijay Kumar Srivastava

We present this book as a guideline to help the graduate students, researchers, and young scientists in the area of biomedical sciences. This book includes the molecular and drug discovery aspects of tropical diseases. The text is presented in a way that is easily accessible to anyone, independent of prior technological knowledge. It is suitable as a reference book for undergraduates and postgraduates training in molecular aspects of drug discovery in tropical disease area along with future prospects to generate research ideas. It can also serve as an introductory book for those pursuing a postgraduate career in health and drug discovery. Those with a background in host and pathogen biology should find this book a valuable addition to the diverse applications of technology in health along with a summary of unique challenges in this domain. This book is divided into six chapters dealing with different human pathogenic tropical diseases, namely Amebiasis, Leishmaniasis, Tuberculosis, Lymphatic filariasis, Malaria, and Sleeping Sickness. Each chapter is a contribution from experts in the field thus enhancing the quality of the content and keeping the molecular aspects of pathology, diagnostics of selected neglected tropical diseases (NTDs). Each chapter begins with an introduction that guides the reader describing the structure, composition, and application of the chapter with the help of selective examples, tables, figures, etc. The conclusion at the end of each chapter summarizes the key points and includes suggestions, wherever required. However, an expert looking for a detailed explanation of any topic pertaining to these tropical diseases is advised to read further from suggested references given at the end of each chapter. We firmly believe that this book provides a valuable resource guide for students to plan and execute their own research on the NTDs with a clear and simple understanding developed after reading this book.

Acknowledgment

A book is incomplete without the efforts of several people who contribute and support in various roles and capacities. We wish to extend our sincere thanks to all the contributing authors for their persistent efforts to bring the best for the enrichment of the target audience. While efforts of some are explicit and visible, some of them have consistently supported our work in background. Gauri takes this opportunity to extend her indebtedness to her mother Mrs. Kamla Misra, her spiritual teacher Late Bhaktivedanta Shri Narayan Swami Maharaj, her grandmother Late Mrs. Shantidevi Misra, and her grandfather Late Mr. Anand Swaroop Misra for their persistent efforts in supporting her shape the career and life. The love and support of all teachers particularly her school Principal Sister Betty Teresa, friends, and colleagues for providing a strong moral foundation have been instrumental in the successful completion of this fourth book. Gauri acknowledges the funding support from ICMR that provides a boost to her scientific and academic endeavours.

Vijay takes this opportunity to extend his heartfelt thanks to his wife Mrs. Chandini Srivastava for her continuous support and affection that has been a constant boost in his academic journey, his sister-in-law Ms. Richa Srivastava, his brother Mr. Ajay Kumar Srivastava, his mother-in-law Mrs. Anjali Nigam, his father-in-law Mr. Naresh Srivastava, his mother Late Mrs. Rama Rani Srivastava, and his father Late Mr. Jitendra Nath Verma for their love, care, constant support, and freedom of choice to choose and shape his career in science and face the challenges of life, emerging as a strong person. The love and support of all teachers, friends, and colleagues have translated this work to a satisfactory conclusion. Sincere thanks to all the reviewers for their valuable feedback that has certainly added weight to the enrichment of this project. The constant support from all the concerned Elsevier officials associated with this project is deeply acknowledged.

Hope the future opens more such enriching possibilities that pave way for academically empowering efforts for the benefit of scientific community.

Amebiasis

Nurulhasanah Othman, Jorim Anak Ujang, Yee Ling Ng, Gaayathri Kumarasamy, Rahmah Noordin

Institute for Research in Molecular Medicine (INFORMM), Universiti Sains Malaysia, Gelugor, Penang, Malaysia

1.1 Introduction

1.1.1 History

Amebiasis was first reported as a deadly disease in 1873 by Hippocrates who examined a patient suffering from bloody dysentery [1]. Two years later, *Entamoeba histolytica* trophozoite was identified by Fedor Aleksondrovich Losch in a farmer who suffered from a fatal case of dysentery [2]. Further investigation by inoculating the stool of the patient into the rectum of a dog caused a similar manifestation [2]. A significant milestone was achieved with the characterization of *E. histolytica* as the causative agent for amebic colitis and amebic liver abscess (ALA) in the 1890s by Sir William Olser and his colleagues [1]. Subsequently, the identification of cyst as an infectious stage was confirmed by Walker and Sellards in 1913 and followed by the establishment of the *E. histolytica* life cycle by Dobell in 1925 [1]. In 1997, amebiasis was ranked second as death-causing parasitic infection, after malaria [3]. Approximately 40,000−100,000 deaths occurred annually, which include 1.9%−9% of amebic colitis patients [4]. The highest morbidity and mortality cases were recorded in tropical and subtropical countries, such as Vietnam, Mexico, and India, where personal hygiene and sanitation are often neglected.

1.1.2 Symptoms

Most patients infected with *E. histolytica* are asymptomatic or only suffered from mild diarrhea [5]. Patients with symptomatic amebiasis often suffer from amebic colitis and ALA [6]. Meanwhile, only 10% of the patients presented classic amebic symptoms such as stomach cramps and bloody diarrhea [7]. Most asymptomatic patients excrete cysts for a short period and are clear from the infection within 12 months of infection [8]. Death occurrence in ALA cases has decreased to 1%−3% due to effective medical intervention. Nonetheless, the mortality rate caused by the late detection resulting in the sudden intraperitoneal rupture occurred in 2%−7% of the patients [6].

Patients with amebic colitis commonly present a history of persistent abdominal pain and diarrhea with the presence of blood and mucus in the stool. As amebiasis is often neglected, a study reports that common inappropriate symptomatic treatment using corticosteroid has led to toxic megacolon complications in about 0.5% of patients [9]. Furthermore, when left untreated, the resulting gut perforation, exhaustion, and extraintestinal amebiasis will lead to death [5]. ALA is the most common extraintestinal manifestation of amebiasis [8]. As mentioned by Zurauskas and McBride [10], patients who develop ALA are usually presented within 5 months of exposure to the disease, with clinical symptoms such as fever and right upper abdominal quadrant pain. Most ALA patients do not present amebic colitis symptoms and also cysts and trophozoites are rarely found in their stools [11].

E. histolytica was not immediately associated as the causative agent of amebiasis because most amebic infections cases were asymptomatic. However, subsequent studies found that the infectious and the noninfectious ameba were not similar [11]. Since then, E. histolytica was reclassified into two species namely the infectious species, E. histolytica and the noninfectious species, E. dispar [11]. Other ameba species such as E. dispar and E. moshkovskii share the same physical features with E. histolytica, thus causing difficulty in differentiating them from E. histolytica under the microscope [3,11]. The need to distinguish E. histolytica from other nonpathogenic Entamoeba species is important to avoid misdiagnosis and wrong treatment.

1.1.3 Lifecycle stages

E. histolytica exists in two distinct stages namely trophozoites and cysts (Fig. 1.1). The simple life cycle begins with the consumption of the tainted fluid containing E. histolytica cyst [5]. The cyst form is round and is enclosed within a refractile wall that protects them from harsh conditions such as stomach acid. It is responsible for the transmission of the disease. A mature cyst is 10−15 μm in size and consists of four nuclei. The cyst withstands harsh environments such as gastric acid. Upon reaching a conducive environment such as the small intestine, a single cyst ex-cyst to form eight trophozoites. The trophozoite form of E. histolytica is 10−50 μm in size and consists of one nucleus. It is actively motile with finger-shaped pseudopodia and responsible for tissue invasion and damage. These blood ingesting trophozoites then colonize the colon and cause dysentery. It is an endoparasite whereby it ingests nutrients from the host and can alter its shape for various purposes such as locomotion and evasion of the host immune responses [12,13]. Trophozoites are unable to live in an unconducive environment outside the host or the host's gastric acids unless quadrinucleate cysts are formed again through a process known as encystation. Humans and primates are the only natural hosts for E. histolytica [6,14]. Despite the advancement in many aspects of the management of amebiasis, the disease remains prevalent in underdeveloped countries of warmer climate [15]. Furthermore, the combination of poor sanitation and bad water quality provides the optimum breeding ground for this parasite [15].

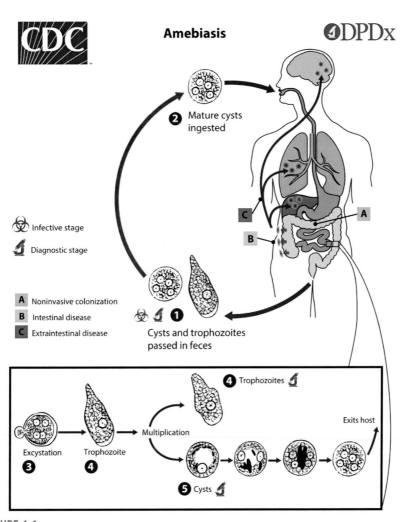

FIGURE 1.1

The life-cycle of *E. histolytica.*

1.2 **Detection of infection**

The earliest diagnosis method of amebiasis is the microscopic examination of stool samples whereby *E. histolytica* trophozoites can be seen containing red blood cells. However, this method is prone to cause misdiagnosis as other morphological similar strains, such as *E. dispar* and *E. moshkovskii*, are indistinguishable from *E. histolytica* under the microscope [16,17]. Although the microscopic method is routinely being used to diagnose amebic colitis, it is not suitable to be performed for the diagnosis of ALA cases. Despite the low sensitivity of microscopy, it is still being practiced in many hospital laboratories.

Amebic colitis patients can also be diagnosed by detecting small ulcers on colonic lesions obtained during the colonoscopic biopsy [18]. Furthermore, colonoscopy and subsequent sampling by means of culture swap are useful in patients with acute colitis and in cases when E. histolytica infection is suspected but failed to be detected in stool samples. However, these methods are time consuming and the sensitivity of the diagnosis is only 50% [19]. Antigen detection methods, for example, Entamoeba CELISA Path kit (Cellabs, Sydney, NSW) and the E. histolytica II kit (TechLab Inc, Blacksburg, VA), are specific and can distinguish E. histolytica from E. dispar. The sensitivities and specificities of these various antigen detection kits range from 80% to 99% and from 86% to 98%, respectively [20−23]. These tests are rapid, and their interpretations are more definitive compared to the microscopic examination.

For the diagnosis of extraintestinal amebiasis such as ALA, radiology imaging is used to detect the presence of an abscess in the liver. When the abscess is present, further analyses such as culture, DNA detection, and/or antigen detection are performed. DNA and antigen detection-based methods performed on the abscess sample were reported to be highly sensitive [1,11,24]. With serological methods, serum samples were used to detect antibodies against E. histolytica for the diagnosis of ALA. Commercial antibody detection assays made of native E. histolytica trophozoite antigens are available [25,26]. However, this method is ineffective to distinguish recent infection from past infection as high background antibody titer may persist in a population of endemic areas though this assay is user independent and more sensitive compared to microscopy and preferred for diagnosis [27−30].

Other techniques such as stool PCR and stool culture are also utilized in many studies. Detection of E. histolytica specific DNA through PCR methods offer high sensitivity and specificity for amebiasis diagnosis. Several types of PCR-based tests have been reported, for example, the real-time PCR, multiplex PCR, conventional PCR, and Loop-Mediated Isothermal Amplification Assay [31−57]. However, a control for PCR is needed to avoid the false-negative result that may be due to the presence of PCR inhibitors in the stool. This method could also be beneficial in the diagnosis of other, unpredicted, problems such as pseudomembranous colitis or inflammatory bowel disease [6]. On the other hand, the application of real-time PCR (RT-PCR) has significantly shortened the detection time by simultaneous monitoring of the amplification process [48]. The advantages of RT-PCR are the ability to detect a low number of parasites and the reliability in differentiating nonpathogenic Entamoeba species from E. histolytica [8]. However, these methods require skilled personnel and the high cost of reagents and equipment.

1.3 Epidemiology and risk factors

E. histolytica infection poses a substantial health risk worldwide, where it affects 40−50 million of people and causes 100,000 deaths per annum globally. Amebiasis commonly occurs in populations living in tropical areas that lack proper sanitation.

The highly endemic countries that have been infected by the invasive diseases caused by *E. histolytica* include Africa [58,59], India and Bangladesh [60], Southeast Asia [61], the Americas [62], and Egypt [63]. The highest morbidity and mortality cases were recorded in Central and South America, Africa, and India [9]. In western Nepal, amebic infection was ranked second after giardiasis [64]. The prevalence of *E. histolytica* infection in the different regions of Brazil from the year 2001 to 2014 ranged between 6.8% and 46.3% [65]. In Pakistan, the prevalence of *E. histolytica* was reported to be as high as 23.1%, whereby the most susceptible age group was found between 6 and 10 years old [66]. The distribution of the disease caused by the parasite from the Africa nations was also assessed through several studies [67,68]. The prevalence of individuals with parasites carrier was ranging from 6% to 75%. However, in Southeast Asian countries like central Vietnam, the incident rate of ALA was recorded as high as 21 cases per 100,000 inhabitants [61].

Amebiasis continues to be a major health problem especially among the aboriginals and communities living in the remote areas of Malaysia [69]. Several studies [70−72] had reported that amebiasis caused by Entamoeba species were prevalent typically in rural areas that had poor socioeconomic conditions, poor environmental sanitation, and lack of personal hygiene practices compared to urban areas. One of the leading factors causing high intestinal parasitic infections including *E. histolytica* in remote communities involves water sources [73]. For instance, the transmission of parasites occurs in settings where a river contaminated with human and animal excretion is used interchangeably for agriculture, socioeconomic, and personal hygiene [73].

Despite being prevalent in tropical countries, human-to-human transmission can still occur regardless of climate and high sanitation standards. For example, in a temperate country such as Japan, mass *E. histolytica* infection at an institution for the mentally disabled in the Yamagata Prefecture of Japan reported 5%−10% of people infected were symptomatic, while 90%−95% of infected subjects were asymptomatic [74]. In a previous report, an amebiasis outbreak in the Netherlands demonstrated that *E. histolytica* can remain dormant for 13 years in their climate [75].

Aspects such as overpopulation, poverty, poor education, and poor hygiene with polluted water supply and unclean conditions contribute to the transmission of amebiasis [76]. Typically, amebiasis is acquired through the fecal-oral route, whereby food or water contaminated by the cyst form of *E. histolytica* is ingested by the host. Human is the only host of amebic infection because there is currently no zoonotic reservoir of *E. histolytica* [77]. Furthermore, the transmission of amebiasis does not involve any insect vectors or other parasites [77]. Transmission usually occurs when there is a direct contact between an infected person and the consumption of cyst-contaminated food or water. Water source plays a vital role in the transmission of amebiasis. In rural areas, the usage of untreated river water among the rural population is commonly practiced. Apart from that, rural communities tend to defecate in the shrubberies and nearby rivers from their houses. This will increase the risk of water pollution and contamination of infectious cysts from human feces mainly during the rainy season. This is because the rainwater will flow down to the ground surface through streams and the contaminated water reaches the rivers, which is unfit for consumption [78]. According to Hankenson et al. [5], the communicability of the disease is high as asymptomatic carriers can be a source of further infection. Furthermore, common household pests like flies and cockroaches can help spread the cyst form of *E. histolytica*. Adults and infants have similar chances of acquiring amebiasis.

Besides, behaviors such as oral-anal sex and homosexual relations can intensify the possibility of getting *E. histolytica* infections [79]. Over past decades, the increased risk of amebiasis in East Asian countries such as Japan, South Korea, and Taiwan has been reported, especially among homosexuals may be due to oral−anal sexual contact [79,80]. According to a report by Hung, Chang, and Ji [80], men who have sex with men have a higher risk of being infected with *E. histolytica*. Apart from the earlier conditions, *E. histolytica* is seldom transmitted in industrialized countries. The outbreaks are very rare in such places [81,82].

Proper health education and community awareness regarding amebiasis should be given to the communities to prevent amebic infection [82]. Practical control measures among households such as appropriate water treatment should be implemented to minimize the potential risk of parasitic ameba cyst-contaminated drinking water. For example, by educating the communities to consume boiled water instead of drinking water from a river or other water sources without boiling.

The global amebiasis epidemiology statistics data remain unclear due to the complication in distinguishing the pathogenic *E. histolytica*, and the nonpathogenic *E. dispar* and *E. moshkovskii*, which are morphologically similar but genetically different [83]. Therefore, instead of relying on the routine diagnostic method in tropical countries such as microscopy, a more specific molecular method namely polymerase chain reaction-based approach was used to differentiate these morphologically identical species [11,83,84].

1.4 Treatment

The first potent tissue amebicide was introduced in 1912, namely emetine. However, this drug causes serious side effects such as vomiting and cardiotoxic [6]. Alternatively, the nitroimidazole derivatives such as metronidazole, tinidazole, and ornidazole are the recommended drugs for amebiasis treatment as they are remarkably safe when compared to emetine [6]. The amebicidal properties of metronidazole were acknowledged and used as the drug for amebiasis treatment in the mid-1960s [85]. Furthermore, metronidazole is also commonly used to treat the infection that is caused by a wide range of anaerobic bacteria effectively due to the low-redox activating enzymes [86,87]. Generally, metronidazole was commonly used in therapeutic and prophylactic for the treatment of major and minor amebiasis that exposed to *E. histolytica*.

The amebicides can be subdivided into two groups: the first-line agents are the 5-nitroimidazole compounds: metronidazole, ornidazole, secnidazole, and tinidazole. In contrast, the second-line agents are emetine hydrochloride, dehydroemetine, and chloroquine phosphate. The treatment of amebiasis using metronidazole often subsequent with luminal agents such as paromomycin and iodoquinol to eradicate the infection, particularly for individuals that are suffered from amebic colitis and amebic liver abscess [86]. Meanwhile, asymptomatic carriers should be treated with a luminal agent to reduce the spread of disease and the risk of developing symptomatic infection [86,88]. According to WHO, it is recommended to treat symptomatic and asymptomatic *E. histolytica* infected individuals [88]. Although metronidazole and other drugs are considered safe to use, the prolonged use of the drug in the infected individuals will cause side effects for instance headache, nausea, vomiting, and abdominal discomfort [85].

For patients with invasive amebiasis, surgical drainage may be unnecessary to treat ALA, as drug therapy alone is efficient. However, aspiration of the abscess was shown to be beneficial in patients with large abscesses [89]. There are variations of drugs that are effective against amebiasis. These drugs can be categorized into two groups. The tissue amebicides work in the liver, bowel, and other places of invasive amebiasis. However, it is not efficient for the treatment of the *E. histolytica* residing in the bowel [87,90]. The recent drug to treat *E. histolytica* and *E. dispar* infection is nitazoxanide. Nitazoxanide could treat 69%−96% of amebiasis cases [91−94]. Asymptomatic infection in children can be treated with diloxanide [95].

1.5 The new strategy for developing next-generation drugs, vaccines, and diagnostic

To date, only a few drugs available and no effective vaccine against amebiasis [96,97]. Hence, the future direction to combat the disease is by developing new antiamebic and vaccine. To achieve this goal, a deep understanding of the mechanism underlying the pathogenesis of this disease is very essential for the elucidation of biological markers that can be explored as drug targets for the development of new drug and vaccine candidates. One such way is by understanding the life-cycle of this parasite that relies on two important forms, that is, trophozoites and cysts that are responsible for amebiasis and transmission, respectively. Human gets infected with *E. histolytica* cyst via the fecal−oral route. Then, the cyst undergoes excystation to form active trophozoites in the small intestine. The trophozoites move, colonize, and proliferate in the large intestine. The proliferated trophozoites eventually develop invasive clinical manifestations such as amebic colitis and ALA. Some of the proliferating trophozoites undergo encystation and form back the cyst. The newly formed cysts and trophozoites are excreted together during the bowel movement. Based on its life cycle, the drug and vaccine that can block the transmission would be an ideal strategy to control the endemicity of amebiasis [98]. These kinds of drugs and vaccines perhaps will be able to cure asymptomatic carriers as they are the source of infections. Thus, it is essential to understand the mechanism of how *E. histolytica* trophozoites undergo encystation to capture biological molecules that can be utilized as drug and vaccine targets?

Mi-Ichi et al. [98] have reviewed a few protein molecules that were found to be involved in encystation that included Gal/GalNac lectin, catecholamine, adrenergic receptor, cholesteryl sulfotransferase, ATP sulfurylase, Ksp90, chitinase, proteasome subunit β-type 5, ubiquitin, and enolase. These protein molecules were suggested to be used as new drug targets. However, due to the information gap of some reported proteins such as enolase, the future study needs to be conducted to further understand the role of these proteins in encystation.

On the other hand, the vaccine and drug that can cure amebiasis by eliminating trophozoites from infected patients is another possible strategy. Previous studies have investigated the protein molecules of trophozoites that are responsible to be involved in the *E. histolytica* pathogenesis, and some of the molecules were investigated as vaccine candidates and drug targets [99−103]. On top of that, several *E. histolytica* membrane proteins are the front line for invasion and evading immune response; hence, they are promising targets for new drugs and vaccines [97,104]. For example, a recent study by Ng et al. [104] has revealed several differentially abundant membrane proteins from the virulent variant of *E. histolytica* including thioredoxin, NAD(P) transhydrogenase subunit alpha, calreticulin, β-N-acetylhexosaminidase, and dipeptidyl-peptidase [104]. These proteins are found to be involved in metabolic and catalytic processes. Therefore, they are potential targets for new drugs and vaccines against amebiasis. The protein, such as EhRabX3 comprising two GTPase domains, is also suggested to be a promising drug target for *E. histolytica* [105].

The establishment of immunological memory is one of the main requirements for effective vaccine development. Besides, an effective vaccine also depends on a strong immune response, the use of an appropriate delivery route and the identification of a defensive antigen [97]. Despite showing immunostimulatory, the use of adjuvants is essential to produce strong antibody. Some of the antigens used as vaccine candidates for amebiasis are shown in Table 1.1.

For example, *E. histolytica* 29-kDa antigen (Eh29) is one of the main target proteins for the amebiasis vaccine. Eh29 is an alkyl hydroperoxide reductase. It is involved in the detoxification of reactive oxygen species (ROS) that is secreted by the immune cells [106]. The previous study has reported 54% of hamsters vaccinated with Eh29 were protected against ALA. Eh29 is a promising vaccine candidate against amebic infections. However, the best mode of delivery or adjuvant is not found. This aspect is important because a successful vaccine depends on the administration route chosen and the type of antigen. Thus, more vaccination trials are needed to find the optimum vaccine treatment suitable for animal models.

Amebiasis outbreak remains in areas that lack proper sanitation systems mainly in developing countries. Therefore, an affordable diagnostic tool without compromises in the sensitivity and specificity would be an ideal solution for those countries. Several diagnostic tools are available in the market ranging from microscopy, ELISA, rapid

Table 1.1 *E. histolytica* antigens tested in animal models.

Antigen	Animal model	Adjuvant/ delivery	Route	% Protection	Immune response	References
SReHP fused to maltose-binding protein	Gerbil	Attenuated *Salmonella typhimurium*	Orally	78% are protected against ALA	Humoral immunity: Anti-SReHP serum IgG, serum IgA, and mucosal IgA	[123]
DNA encoding SReHP	Mouse Gerbil	Plasmid	Intramusculary	Mouse: 80% are protected against ALA Gerbils: 60% are protected against ALA	Humoral immunity: anti-SReHP IgG strong lymphocyte proliferation	[124]
29-kDa alkyl hydroperoxide reductase	Mice	CTB	Orally	80% are protected against AC	Humoral immunity: Anti-eh29 intestinal IgA and serum IgG	[125]
DNA encoding ehcp112 and ehadh112	Hamster	Plasmid	Intradermally intramusculary	i.d.: 60% survived	Poor humoral immunity for both delivery routes (only IgG measured)	[126]
HSBP	Guinea pig	Freund's	Subcutaneously	N/A	Humoral immunity: anti-HSBP IgG and IgM, IgA	[127]

%Protection: ([number of uninfected vaccinated animals]/[total number of vaccinated animals]) × 100; AC, amebic colitis; ALA, amebic liver abscess.

test to molecular detection. However, limitations such as low sensitivity of microscopy, higher cost of molecular detection, and variable sensitivities and specificities of serology tests are among major challenges in the diagnosis of amebiasis mainly for the endemic areas [107]. The discovery of a new biological marker is highly needed using the combination of "omics" technologies for amebiasis diagnosis. An ideal biological marker would be a molecule that can differentiate infection stages and nonpathogenic Entamoeba. In addition, the point-of-care or rapid test is suggested to be the best diagnostic format for routine and field-applicable tests due to its simplicity.

1.6 New pathways that can be suitable drug targets for disease intervention

Recent studies have suggested several putative molecular pathways including de novo L-cysteine biosynthesis, Co A biosynthetic pathway, actin cytoskeleton remodeling, stress response, and removal of acetyl groups from histone and nonhistone proteins as new drug targets [108−112]. As such, a few inhibitors have been tested to target earlier pathways such as Pencolide, Teicoplanin, Src inhibitor, Rutilantin, and Vorinostat.

Using molecular techniques to study epigenetic and RNA silencing in *E. histolytica*, several new pathways that involve genes encode for Dephospho-CoA kinase, EhRrp6 exoribonuclease, HP127670 uncharacterized gene, and AIG1 (EHI_176590) involve in the formation of surface protrusion are also suggested as new drug targets [113−116].

In-depth study of *E. histolytica* acetyl Co-A pathway that consists of four main enzymes, which are pantothenate kinase (PanK, EC 2.7.1.33), bifunctional phosphopantothenate-cysteine ligase/decarboxylase, phosphopantetheine adenylyltransferase, and dephospho-CoA kinase, is previously studied. Due to the homolog divergent of *E. histolytica* PanK to human orthologue and its physiological importance, this enzyme was selected for the development of a new inhibitor target [108]. Screening results of compounds against recombinant PanK from Kitasato Natural Products Library revealed that 14 compounds showed inhibition activities. The best compound showed moderate inhibition of PanK activity and cell growth at a low concentration, as well as differential toxicity toward *E. histolytica* and human cells.

On the other hand, thioredoxin reductase that is involved in protecting *E. histolytica* from ROS during the tissue invasion would make this molecule suitable for the drug target. Previous studies have shown that the thioredoxin reductase (EhTrxR/Trx) system reduces metronidazole [117]. This enzyme also interacts with downstream peroxidases that are critical for cellular redox homeostasis [118]. Despite metronidazole, auronofin was also investigated in targeting EhTrxR and showed promising results highlighted antiamebic activities against *E. histolytica* trophozoites [101]. Further study involving clinical trials will be conducted on auronofin to validate its performance [103].

Natural compounds such as flavonoids also have been investigated as antiamebic agents [119]. Several studies showed (−)-epicatechin targeted on alteration of *E histolytica* nucleus and actin [120], while Kaempferol and Tiliroside inhibited pyruvate:ferredoxin oxidoreductase-PFOR and fructose-1,6-bisphosphate aldolase -G/FBPA [121]. Resveratrol (Polyphenol) is targeting cell growth arrested, generation of oxidative stress, and damage cell membrane lipids in vivo [122].

Acknowledgment

We like to acknowledge Universiti Sains Malaysia for the financial and facility supports.

References

[1] Tanyuksel M, Petri WA. Laboratory diagnosis of amebiasis. Clinical Microbiology Reviews 2003;16(4):713−29.

[2] Marshall MM, Naumovitz D, Ortega Y, Sterling CR. Waterborne protozoan pathogens. Clinical Microbiology Reviews 1997;10(1):67−85.

[3] Anonymous. WHO/PAHO/UNESCO report. A consultation with experts on amoebiasis. Epidemiological Bulletin/PAHO 1997;18:3−14.

[4] Aristizábal H, Acevedo J, Botero M. Fulminant amebic colitis. World Journal of Surgery 1991;15(2):216−21.

[5] Hankenson FC, Johnston NA, Weigler BJ, Di Giacomo RF. Zoonoses of occupational health importance in contemporary laboratory animal research. Comparative Medicine 2003;53(6):579−601.

[6] Stanley Jr SL. Amoebiasis. The Lancet 2003;361(9362):1025−34.

[7] Farthing MJ. Treatment options for the eradication of intestinal protozoa. Nature Reviews Gastroenterology and Hepatology 2006;3(8):436.

[8] van Hal SJ, Stark DJ, Fotedar R, Marriott D, Ellis JT, Harkness JL. Amoebiasis: current status in Australia. Medical Journal of Australia 2007;186(8):412.

[9] Ackers J, Clark C, Diamond L, Duchene M, Espinosa-Cantellano M, Jackson T, et al. WHO/PAHO. UNESCO report A consultation with experts on amoebiasis Mexico City, Mexico. 1997. 28:29.

[10] Zurauskas J, McBride W. Case of amoebic liver abscess: prolonged latency or acquired in Australia? Internal Medicine Journal 2001;31(9):565−6.

[11] Fotedar R, Stark D, Beebe N, Marriott D, Ellis J, Harkness J. Laboratory diagnostic techniques for Entamoeba species. Clinical Microbiology Reviews 2007;20(3): 511−32.

[12] Espinosa-Cantellano M, Chavez B, Calderon J, Martinez-Palomo A. *Entamoeba histolytica*: electrophoretic analysis of isolated caps induced by several ligands. Archives of Medical Research 1992;23(2):81−5.

[13] Markiewicz JM, Syan S, Hon C-C, Weber C, Faust D, Guillen N. A proteomic and cellular analysis of uropods in the pathogen *Entamoeba histolytica*. PLoS Neglected Tropical Diseases 2011;5(4):e1002.

[14] Rivera WL, Yason JAD, Adao DEV. *Entamoeba histolytica* and *E. dispar* infections in captive macaques (*Macaca fascicularis*) in the Philippines. Primates 2010;51(1):69.

[15] Kean BH. A history of amebiasis, p. 1−10. In: Ravdin JI, editor. Amebiasis: human infection by Entamoeba histolytica. New York, N.Y: John Wiley & Sons, Inc.; 1988.

[16] Liang S-Y, Chan Y-H, Hsia K-T, Lee J-L, Kuo M-C, Hwa K-Y, et al. Development of loop-mediated isothermal amplification assay for detection of *Entamoeba histolytica*. Journal of Clinical Microbiology 2009;47(6):1892−5.

[17] Haque R, Petri WA. Diagnosis of amebiasis in Bangladesh. Archives of Medical Research 2006;37(2):272−5.

[18] Ohnishi K, Kato Y, Imamura A, Fukayama M, Tsunoda T, Sakaue Y, et al. Present characteristics of symptomatic *Entamoeba histolytica* infection in the big cities of Japan. Epidemiology and Infection 2004;132(1):57−60.

[19] Clark CG, Diamond LS. Methods for cultivation of luminal parasitic protists of clinical importance. Clinical Microbiology Reviews 2002;15(3):329−41.

[20] Haque R, Neville LM, Hahn P, Petri W. Rapid diagnosis of *Entamoeba* infection by using *Entamoeba* and *Entamoeba histolytica* stool antigen detection kits. Journal of Clinical Microbiology 1995;33(10):2558−61.

[21] Gonin P, Trudel L. Detection and differentiation of *Entamoeba histolytica* and *Entamoeba dispar* isolates in clinical samples by PCR and enzyme-linked immunosorbent assay. Journal of Clinical Microbiology 2003;41(1):237−41.

[22] Furrows S, Moody A, Chiodini P. Comparison of PCR and antigen detection methods for diagnosis of *Entamoeba histolytica* infection. Journal of Clinical Pathology 2004; 57(12):1264−6.

[23] Solaymani-Mohammadi S, Rezaian M, Babaei Z, Rajabpour A, Meamar AR, Pourbabai AA, et al. Comparison of a stool antigen detection kit and PCR for diagnosis of *Entamoeba histolytica* and *Entamoeba dispar* infections in asymptomatic cyst passers in Iran. Journal of Clinical Microbiology 2006;44(6):2258−61.

[24] Paul J, Srivastava S, Bhattacharya S. Molecular methods for diagnosis of *Entamoeba histolytica* in a clinical setting: an overview. Experimental Parasitology 2007;116(1): 35−43.

[25] Lotter H, Mannweiler E, Schreiber M, Tannich E. Sensitive and specific serodiagnosis of invasive amebiasis by using a recombinant surface protein of pathogenic *Entamoeba histolytica*. Journal of Clinical Microbiology 1992;30(12):3163−7.

[26] Ning TZ, Kin WW, Noordin R, Cun STW, Chong FP, Mohamed Z, et al. Evaluation of *Entamoeba histolytica* recombinant phosphoglucomutase protein for serodiagnosis of amoebic liver abscess. BMC Infectious Diseases 2013;13(1):144.

[27] Pillai DR, Keystone JS, Sheppard DC, MacLean JD, MacPherson DW, Kain KC. *Entamoeba histolytica* and *Entamoeba dispar*: epidemiology and comparison of diagnostic methods in a setting of nonendemicity. Clinical Infectious Diseases 1999;29(5): 1315−8.

[28] Zengzhu G, Bracha R, Nuchamowitz Y, Cheng W, Mirelman D. Analysis by enzymelinked immunosorbent assay and PCR of human liver abscess aspirates from patients in China for *Entamoeba histolytica*. Journal of Clinical Microbiology 1999;37(9): 3034−6.

[29] Zeehaida M, WNA WA, Amry A, Hassan S, Sarimah A, Rahmah N. A study on the usefulness of Techlab *Entamoeba histolytica* II antigen detection ELISA in the diagnosis of amoebic liver abscess (ALA) at Hospital Universiti Sains Malaysia (HUSM), Kelantan, Malaysia. Tropical Biomedicine 2008;25(3):209−16.

[30] Mohamed Z, Bachok N, Hasan H. Analysis of indirect hemagglutination assay results among patients with amoebic liver abscess. International Medical Journal 2009;16(3).

[31] Blessmann J, Buss H, Nu PA, Dinh BT, Ngo QT, Van AL, et al. Real-time PCR for detection and differentiation of *Entamoeba histolytica* and *Entamoeba dispar* in fecal samples. Journal of Clinical Microbiology 2002;40(12):4413−7.

[32] Verweij JJ, Vermeer J, Brienen EA, Blotkamp C, Laeijendecker D, van Lieshout L, et al. *Entamoeba histolytica* infections in captive primates. Parasitology Research 2003;90(2):100−3.

[33] Kebede A, Verweij JJ, Endeshaw T, Messele T, Tasew G, Petros B, et al. The use of real-time PCR to identify *Entamoeba histolytica* and *E. dispar* infections in prisoners and primary-school children in Ethiopia. Annals of Tropical Medicine and Parasitology 2004;98(1):43−8.

[34] Verweij JJ, Blangé RA, Templeton K, Schinkel J, Brienen EA, van Rooyen MA, et al. Simultaneous detection of *Entamoeba histolytica*, *Giardia lamblia*, and *Cryptosporidium parvum* in fecal samples by using multiplex real-time PCR. Journal of Clinical Microbiology 2004;42(3):1220−3.

[35] Qvarnstrom Y, James C, Xayavong M, Holloway BP, Visvesvara GS, Sriram R, et al. Comparison of real-time PCR protocols for differential laboratory diagnosis of amebiasis. Journal of Clinical Microbiology 2005;43(11):5491−7.

[36] Roy S, Kabir M, Mondal D, Ali IK, Petri WA, Haque R. Real-time-PCR assay for diagnosis of *Entamoeba histolytica* infection. Journal of Clinical Microbiology 2005;43(5):2168−72.

[37] Calderaro A, Gorrini C, Bommezzadri S, Piccolo G, Dettori G, Chezzi C. *Entamoeba histolytica* and *Entamoeba dispar*: comparison of two PCR assays for diagnosis in a non-endemic setting. Transactions of the Royal Society of Tropical Medicine and Hygiene 2006;100(5):450−7.

[38] Visser LG, Verweij JJ, Van Esbroeck M, Edeling WM, Clerinx J, Polderman AM. Diagnostic methods for differentiation of *Entamoeba histolytica* and *Entamoeba dispar* in carriers: performance and clinical implications in a non-endemic setting. International Journal of Medical Microbiology 2006;296(6):397−403.

[39] Ahmad N, Khan M, Hoque MI, Haque R, Mondol D. Detection of *Entamoeba histolytica* DNA from liver abscess aspirate using polymerase chain reaction (PCR): a diagnostic tool for amoebic liver abscess. Bangladesh Medical Research Council Bulletin 2007;33(1):13−20.

[40] Haque R, Roy S, Siddique A, Mondal U, Rahman SM, Mondal D, et al. Multiplex real-time PCR assay for detection of *Entamoeba histolytica*, *Giardia intestinalis*, and *Cryptosporidium* spp. The American Journal of Tropical Medicine and Hygiene 2007;76(4):713−7.

[41] ten Hove R, Schuurman T, Kooistra M, Möller L, van Lieshout L, Verweij JJ. Detection of diarrhoea-causing protozoa in general practice patients in the Netherlands by multiplex real-time PCR. Clinical Microbiology and Infections 2007;13(10):1001−7.

[42] Singh A, Houpt E, Petri WA. Rapid diagnosis of intestinal parasitic protozoa, with a focus on *Entamoeba histolytica*. Interdisciplinary Perspectives on Infectious Diseases 2009;2009:547090.

[43] ten Hove RJ, van Esbroeck M, Vervoort T, van den Ende J, van Lieshout L, Verweij JJ. Molecular diagnostics of intestinal parasites in returning travellers. European Journal of Clinical Microbiology and Infectious Diseases 2009;28(9):1045−53.

[44] Gutiérrez-Cisneros MJ, Cogollos R, López-Vélez R, Martín-Rabadán P, Martínez-Ruiz R, Subirats M, et al. Application of real-time PCR for the differentiation of *Entamoeba histolytica* and *E. dispar* in cyst-positive faecal samples from 130 immigrants living in Spain. Annals of Tropical Medicine and Parasitology 2010; 104(2):145−9.

[45] Hamzah Z, Petmitr S, Mungthin M, Leelayoova S, Chavalitshewinkoon-Petmitr P. Development of multiplex real-time polymerase chain reaction for detection of *Entamoeba histolytica*, *Entamoeba dispar*, and *Entamoeba moshkovskii* in clinical specimens. The American Journal of Tropical Medicine and Hygiene 2010;83(4): 909−13.

[46] Haque R, Kabir M, Noor Z, Rahman SM, Mondal D, Alam F, et al. Diagnosis of amebic liver abscess and amebic colitis by detection of *Entamoeba histolytica* DNA in blood, urine, and saliva by a real-time PCR assay. Journal of Clinical Microbiology 2010;48(8):2798−801.

[47] Liang SY, Hsia KT, Chan YH, Fan CK, Jiang DD, Landt O, et al. Evaluation of a new single-tube multiprobe real-time PCR for diagnosis of *Entamoeba histolytica* and *Entamoeba dispar*. The Journal of Parasitology 2010;96(4):793−7.

[48] Othman N, Mohamed Z, Verweij JJ, Huat LB, Olivos-García A, Yeng C, et al. Application of real-time polymerase chain reaction in detection of *Entamoeba histolytica* in pus aspirates of liver abscess patients. Foodborne Pathogens and Disease 2010;7(6):637−41.

[49] Santos HL, Bandyopadhyay K, Bandea R, Peralta RH, Peralta JM, Da Silva AJ. LUMINEX®: a new technology for the simultaneous identification of five *Entamoeba* spp. commonly found in human stools. Parasites and Vectors 2013;6:69.

[50] Van Lint P, Rossen JW, Vermeiren S, Ver Elst K, Weekx S, Van Schaeren J, et al. Detection of *Giardia lamblia*, *Cryptosporidium* spp. and *Entamoeba histolytica* in clinical stool samples by using multiplex real-time PCR after automated DNA isolation. Acta Clinica Belgica 2013;68(3):188−92.

[51] Zebardast N, Yeganeh F, Gharavi MJ, Abadi A, Seyyed Tabaei SJ, Haghighi A. Simultaneous detection and differentiation of *Entamoeba histolytica*, *E. dispar*, *E. moshkovskii*, *Giardia lamblia* and *Cryptosporidium* spp. in human fecal samples using multiplex PCR and qPCR-MCA. Acta Tropica 2016;162:233−8.

[52] Mero S, Kirveskari J, Antikainen J, Ursing J, Rombo L, Kofoed PE, et al. Multiplex PCR detection of *Cryptosporidium* sp, *Giardia lamblia* and *Entamoeba histolytica* directly from dried stool samples from Guinea-Bissauan children with diarrhoea. Information Display 2017;49(9):655−63.

[53] Morio F, Valot S, Laude A, Desoubeaux G, Argy N, Nourrisson C, et al. Evaluation of a new multiplex PCR assay (ParaGENIE G-Amoeba Real-Time PCR kit) targeting *Giardia intestinalis*, *Entamoeba histolytica* and *Entamoeba dispar*/*Entamoeba moshkovskii* from stool specimens: evidence for the limited performances of microscopy-based approach for amoeba species identification. Clinical Microbiology and Infections 2018;24(11):1205−9.

[54] Parčina M, Reiter-Owona I, Mockenhaupt FP, Vojvoda V, Gahutu JB, Hoerauf A, et al. Highly sensitive and specific detection of *Giardia duodenalis*, *Entamoeba histolytica*, and *Cryptosporidium* spp. in human stool samples by the BD MAX™ Enteric Parasite Panel. Parasitology Research 2018;117(2):447−51.

[55] Schuurs TA, Koelewijn R, Brienen EAT, Kortbeek T, Mank TG, Mulder B, et al. Harmonization of PCR-based detection of intestinal pathogens: experiences from the Dutch external quality assessment scheme on molecular diagnosis of protozoa in stool samples. Clinical Chemistry and Laboratory Medicine 2018;56(10):1722−7.

[56] Chihi A, Stensvold CR, Ben-Abda I, Ben-Romdhane R, Aoun K, Siala E, et al. Development and evaluation of molecular tools for detecting and differentiating intestinal amoebae in healthy individuals. Parasitology 2019:1−7.

[57] Guevara Á, Vicuña Y, Costales D, Vivero S, Anselmi M, Bisoffi Z, et al. Use of real-time polymerase chain reaction to differentiate between pathogenic. The American Journal of Tropical Medicine and Hygiene 2019;100(1):81−2.

[58] Stauffer W, Ravdin JI. *Entamoeba histolytica*: an update. Current Opinion in Infectious Diseases 2003;16(5):479−85.

[59] Abd-Alla MD, Jackson TF, Rogers T, Reddy S, Ravdin JI. Mucosal immunity to asymptomatic *Entamoeba histolytica* and *Entamoeba dispar* infection is associated with a peak intestinal anti-lectin immunoglobulin A antibody response. Infection and Immunity 2006;74(7):3897−903.

[60] Haque R, Ali IM, Sack RB, Farr BM, Ramakrishnan G, Petri WA. Amebiasis and mucosal IgA antibody against the *Entamoeba histolytica* adherence lectin in Bangladeshi children. The Journal of Infectious Diseases 2001;183(12):1787−93.

[61] Blessmann J, Van Linh P, Nu PA, Thi HD, Muller-Myhsok B, Buss H, et al. Epidemiology of amebiasis in a region of high incidence of amebic liver abscess in central Vietnam. The American Journal of Tropical Medicine and Hygiene 2002;66(5): 578−83.

[62] Ramos F, Morán P, González E, García G, Ramiro M, Gómez A, et al. High prevalence rate of *Entamoeba histolytica* asymptomatic infection in a rural Mexican community. The American Journal of Tropical Medicine and Hygiene 2005;73(1):87−91.

[63] Abd-Alla MD, Wahib AA, Ravdin JI. Comparison of antigen-capture ELISA to stool-culture methods for the detection of asymptomatic *Entamoeba* species infection in Kafer Daoud, Egypt. The American Journal of Tropical Medicine and Hygiene 2000;62(5):579−82.

[64] Sharma BK, Rai SK, Rai DR, Choudhury DR. Prevalence of intestinal parasitic infestation in schoolchildren in the northeastern part of Kathmandu Valley, Nepal. Southeast Asian Journal of Tropical Medicine and Public Health 2004;35(3):501−5.

[65] Silva MT, Santana JV, Bragagnoli G, Marinho AM, Malagueño E. Prevalence of *Entamoeba histolytica/Entamoeba dispar* in the city of Campina Grande, in northeastern Brazil. Revista do Instituto de Medicina Tropical de Sao Paulo 2014;56(5): 451−4.

[66] Alam MA, Maqbool A, Nazir MM, Lateef M, Khan MS, Ahmed AN, et al. Prevalence of *Entamoeba histolytica* -like cysts compared to *E. histolytica* antigens detected by ELISA in the stools of 600 patients from three socioeconomic communities in the Metropolitan City of Lahore, Pakistan. The Journal of Parasitology 2015;101(2): 236−9.

[67] Samie A, Obi CL, Lall N, Meyer JJ. In-vitro cytotoxicity and antimicrobial activities, against clinical isolates of *Campylobacter* species and *Entamoeba histolytica*, of local medicinal plants from the Venda region, in South Africa. Annals of Tropical Medicine and Parasitology 2009;103(2):159−70.

[68] Fujii Y, Kaneko S, Nzou SM, Mwau M, Njenga SM, Tanigawa C, et al. Serological surveillance development for tropical infectious diseases using simultaneous microsphere-based multiplex assays and finite mixture models. PLoS Neglected Tropical Diseases 2014;8(7):e3040.

[69] Tengku SA, Norhayati M. Public health and clinical importance of amoebiasis in Malaysia: a review. Tropical Biomedicine 2011;28(2):194−222.

[70] Rajeswari B, Sinniah B, Hussein H. Socio-economic factors associated with intestinal parasites among children living in Gombak, Malaysia. Asia-Pacific Journal of Public Health 1994;7(1):21−5.

[71] Hakim SL, Gan CC, Malkit K, Azian MN, Chong CK, Shaari N, et al. Parasitic infections among Orang Asli (aborigine) in the Cameron Highlands, Malaysia. Southeast Asian Journal of Tropical Medicine and Public Health 2007;38(3):415−9.

[72] Ngui R, Ishak S, Chuen CS, Mahmud R, Lim YA. Prevalence and risk factors of intestinal parasitism in rural and remote West Malaysia. PLoS Neglected Tropical Diseases 2011;5(3):e974.

[73] Pham Duc P, Nguyen-Viet H, Hattendorf J, Zinsstag J, Dac Cam P, Odermatt P. Risk factors for *Entamoeba histolytica* infection in an agricultural community in Hanam province, Vietnam. Parasites and Vectors 2011;4:102.

[74] Haghighi A, Kobayashi S, Takeuchi T, Masuda G, Nozaki T. Remarkable genetic polymorphism among *Entamoeba histolytica* isolates from a limited geographic area. Journal of Clinical Microbiology 2002;40(11):4081−90.

[75] Vreden SG, Visser LG, Verweij JJ, Blotkamp J, Stuiver PC, Aguirre A, et al. Outbreak of amebiasis in a family in the Netherlands. Clinical Infectious Diseases 2000;31(4):1101−4.

[76] Akhtar T, Khan AG, Ahmed I, Nazli R, Haider J. Prevalence of amoebiasis in a model research community and its confirmation using stool antigen elisa for *Entamoeba histolytica*. Pakistan Journal of Pharmaceutical Sciences 2016;29(5):1587−90.

[77] Petri WA. Pathogenesis of amebiasis. Current Opinion in Microbiology 2002;5(4):443−7.

[78] Ngui R, Angal L, Fakhrurrazi SA, Lian YL, Ling LY, Ibrahim J, et al. Differentiating *Entamoeba histolytica*, *Entamoeba dispar* and *Entamoeba moshkovskii* using nested polymerase chain reaction (PCR) in rural communities in Malaysia. Parasites and Vectors 2012;5:187.

[79] Hung CC, Ji DD, Sun HY, Lee YT, Hsu SY, Chang SY, et al. Increased risk for *Entamoeba histolytica* infection and invasive amebiasis in HIV seropositive men who have sex with men in Taiwan. PLoS Neglected Tropical Diseases 2008;2(2):e175.

[80] Hung CC, Chang SY, Ji DD. *Entamoeba histolytica* infection in men who have sex with men. The Lancet Infectious Diseases 2012;12(9):729−36.

[81] Gunther J, Shafir S, Bristow B, Sorvillo F. Short report: amebiasis-related mortality among United States residents, 1990-2007. The American Journal of Tropical Medicine and Hygiene 2011;85(6):1038−40.

[82] Shirley DT, Farr L, Watanabe K, Moonah S. A review of the global burden, new diagnostics, and current therapeutics for amebiasis. Open Forum Infectious Diseases 2018;5(7):ofy161.

[83] Ali IK, Clark CG, Petri WA. Molecular epidemiology of amebiasis. Infection, Genetics and Evolution 2008;8(5):698—707.

[84] Fotedar R, Stark D, Beebe N, Marriott D, Ellis J, Harkness J. PCR detection of *Entamoeba histolytica*, *Entamoeba dispar*, and *Entamoeba moshkovskii* in stool samples from Sydney, Australia. Journal of Clinical Microbiology 2007;45(3):1035—7.

[85] Lucas R, Upcroft JA. Clinical significance of the redefinition of the agent of amoebiasis. Revista Latinoamericana de Microbiología 2001;43(4):183—7.

[86] Freeman CD, Klutman NE, Lamp KC. Metronidazole. A therapeutic review and update. Drugs 1997;54(5):679—708.

[87] Gonzales MLM, Dans LF, Sio-Aguilar J. Antiamoebic drugs for treating amoebic colitis. Cochrane Database of Systematic Reviews 2019;1. CD006085.

[88] Ximénez C, Morán P, Rojas L, Valadez A, Gómez A, Ramiro M, et al. Novelties on amoebiasis: a neglected tropical disease. Journal of Global Infectious Diseases 2011;3(2):166—74.

[89] Cordel H, Prendki V, Madec Y, Houze S, Paris L, Bourée P, et al. Imported amoebic liver abscess in France. PLoS Neglected Tropical Diseases 2013;7(8):e2333.

[90] Gonzales ML, Dans LF, Martinez EG. Antiamoebic drugs for treating amoebic colitis. Cochrane Database of Systematic Reviews 2009;2. CD006085.

[91] Abaza H, El-Zayadi AR, Kabil SM, Rizk H. Nitazoxanide in the treatment of patients with intestinal protozoan and helminthic infections: a report on 546 patients in egypt. Current Therapeutic Research 1998;59(2):116—21. https://doi.org/10.1016/S0011-393X(98)85006-6.

[92] Romero CR, Guerrero LR, Muñóz García MR, Geyne Cruz A. Nitazoxanide for the treatment of intestinal protozoan and helminthic infections in Mexico. Transactions of the Royal Society of Tropical Medicine and Hygiene 1997;91(6):701—3.

[93] Davila-Gutierrez CE, Vasquez C, Trujillo-Hernandez B, Huerta M. Nitazoxanide compared with quinfamide and mebendazole in the treatment of helminthic infections and intestinal protozoa in children. The American Journal of Tropical Medicine and Hygiene 2002;66(3):251—4.

[94] Diaz E, Mondragon J, Ramirez E, Bernal R. Epidemiology and control of intestinal parasites with nitazoxanide in children in Mexico. The American Journal of Tropical Medicine and Hygiene 2003;68(4):384—5.

[95] Escobedo AA, Almirall P, Alfonso M, Cimerman S, Rey S, Terry SL. Treatment of intestinal protozoan infections in children. Archives of Disease in Childhood 2009; 94:478—82. https://doi.org/10.1136/adc.2008.151852.

[96] Haque R, Huston CD, Hughes M, Houpt E, Petri WA. Amebiasis. New England Journal of Medicine 2003;348(16):1565—73.

[97] Quach J, St-Pierre J, Chadee K. The future for vaccine development against *Entamoeba histolytica*. Human Vaccines and Immunotherapeutics 2014;10(6): 1514—21.

[98] Mi-Ichi F, Yoshida H, Hamano S. *Entamoeba encystation*: new targets to prevent the transmission of amebiasis. PLoS Pathogens 2016;12(10):e1005845.

[99] Singh RS, Walia AK, Kanwar JR, Kennedy JF. Amoebiasis vaccine development: a snapshot on *E. histolytica* with emphasis on perspectives of Gal/GalNAc lectin. International Journal of Biological Macromolecules 2016;91:258—68.

[100] Min X, Feng M, Guan Y, Man S, Fu Y, Cheng X, et al. Evaluation of the C-Terminal fragment of *Entamoeba histolytica* Gal/GalNAc lectin intermediate subunit as a vaccine candidate against amebic liver abscess. PLoS Neglected Tropical Diseases 2016;10(1):e0004419.

[101] Debnath A, Parsonage D, Andrade RM, He C, Cobo ER, Hirata K, et al. A high-throughput drug screen for *Entamoeba histolytica* identifies a new lead and target. Nature Medicine 2012;18(6):956−60. https://doi.org/10.1038/nm.2758.

[102] Parsonage D, Sheng F, Hirata K, Debnath A, McKerrow JH, Reed SL, et al. X-ray structures of thioredoxin and thioredoxin reductase from *Entamoeba histolytica* and prevailing hypothesis of the mechanism of Auranofin action. Journal of Structural Biology 2016;194(2):180−90.

[103] Andrade RM, Reed SL. New drug target in protozoan parasites: the role of thioredoxin reductase. Frontiers in Microbiology 2015;6:975. https://doi.org/10.3389/fmicb 2015.00975.

[104] Ng YL, Olivos-García A, Lim TK, Noordin R, Lin Q, Othman N. Entamoeba histolytica: quantitative proteomics analysis reveals putative virulence-associated differentially abundant membrane proteins. American Journal of Tropical Medicine and Hygiene 2018;99(6):1518−29. https://doi.org/10.4269/ajtmh.18-0415.

[105] Srivastava VK, Chandra M, Saito-Nakano Y, Nozaki T, Datta S. Crystal structure analysis of wild type and fast hydrolyzing mutant of EhRabX3, a Tandem ras superfamily GTPase from *Entamoeba histolytica*. Journal of Molecular Biology 2016;428(1): 41−51.

[106] Bruchhaus I, Richter S, Tannich E. Removal of hydrogen peroxide by the 29 kDa protein of *Entamoeba histolytica*. Biochemical Journal 1997;326(Pt 3):785−9.

[107] Saidin S, Othman N, Noordin R. Update on laboratory diagnosis of amoebiasis. European Journal of Clinical Microbiology and Infectious Diseases 2019;38(1):15−38.

[108] Nurkanto A, Jeelani G, Yamamoto T, Naito Y, Hishiki T, Mori M, et al. Characterization and validation of *Entamoeba histolytica* pantothenate kinase as a novel anti-amebic drug target. International Journal for Parasitology: Drugs and Drug Resistance 2018;8(1):125−36.

[109] Jeelani G, Sato D, Soga T, Nozaki T. Genetic, metabolomic and transcriptomic analyses of the de novo L-cysteine biosynthetic pathway in the enteric protozoan parasite *Entamoeba histolytica*. Scientific Reports 2017;7:15649.

[110] Lopez-Contreras L, Hernandez-Ramirez VI, Herrera-Martinez M, Montano S, Constantino-Jonapa LA, Chavez-Munguia B, Talamas-Rohana P. Structural and functional characterization of the divergent i Src using Src inhibitor-1. Parasites and Vectors 2017;10:500.

[111] Shahinas D, Debnath A, Benedict C, McKerrow JH, Pillai DR. Heat shock protein 90 inhibitors repurposed against *Entamoeba histolytica*. Frontiers in Microbiology 2015; 6:368.

[112] Sarita M, Luis Alejandro CJ, Yudibeth SL, Veronica Ivonne HR, Ceruelos Alejandra H, Luis Carlos RQ, Ledezma Jesus Carlos R, Patricia TR, Contreras Luilli L. Vorinostat, a possible alternative to metronidazole for the treatment of of amebiasis caused by *Entamoeba histolytica*. Journal of Biomolecular Structure and Dynamics 2019:1−9.

[113] Nurkanto A, Jeelani G, Yamamoto T, Hishiki T, Naito Y, Suematsu M, Hashimoto T, Nozaki T. Biochemical, metabolomic, and genetic analyses of dephospho coenzyme A kinase involved in coenzyme A biosynthesis in the human enteric parasite *Entamoeba histolytica*. Frontiers in Microbiology 2018;9:2902.

[114] Singh SS, Naiyer S, Bharadwaj R, Kumar A, Singh YP, Ray AK, Subbarao N, Bhattacharya A, Bhattacharya S. Stress-induced nuclear depletion of *Entamoeba histolytica* 3'−5' exoribonuclease EhRrp6 and its role in growth and erythrophagocytosis. Journal of Biological Chemistry 2018;293:16242−60.

[115] Matthiesen J, Lender C, Haferkorn A, Fehling H, Meyer M, Matthies T, Tannich E, Roeder T, Lotter H, Bruchhaus I. Trigger-induced RNAi gene silencing to identify pathogenicity factors of *Entamoeba histolytica*. The FASEB Journal 2019;33: 1658−68.

[116] Nakada-Tsukui K, Sekizuka T, Sato-Ebine E, Escueta-de Cadiz A, Ji DD, Tomii K, Kuroda M, Nozaki T. AIG1 affects in vitro and in vivo virulence in clinical isolates of *Entamoeba histolytica*. PLoS Pathogens 2018;14:e1006882.

[117] Leitsch D, Kolarich D, Wilson IB, Altmann F, Duchêne. Nitroimidazole action in *Entamoeba histolytica*: a central role for thioredoxin reductase. PLoS Biology 2007; 5:e211. https://doi.org/10.1371/journal.pbio.0050211.

[118] Schlosser S, Leitsch D, Duchêne M. *Entamoeba histolytica*: identification of thioredoxin-targeted proteins and analysis of serine acetyltransferase-1 as a prototype example. Biochemical Journal 2013;451:277−88. https://doi.org/10.1042/BJ2012 1798.

[119] Martínez-Castillo M, Pacheco-Yepez J, Flores-Huerta N, Guzmán-Téllez P, Jarillo-Luna RA, Cárdenas-Jaramillo LM, et al. Flavonoids as a natural treatment against. Front Cell Infect Microbiology 2018;8:209.

[120] López-Soto F, León-Sicairos N, Nazmi K, Bolscher JG, de la Garza M. Microbicidal effect of the lactoferrin peptides lactoferricin17-30, lactoferrampin265-284, and lacto-ferrin chimera on the parasite *Entamoeba histolytica*. Biometals 2010;23(3):563−8.

[121] Bolaños V, Díaz-Martínez A, Soto J, Marchat LA, Sanchez-Monroy V, Ramírez-Moreno E. Kaempferol inhibits *Entamoeba histolytica* growth by altering cytoskeletal functions. Molecular and Biochemical Parasitology 2015;204(1):16−25.

[122] Pais-Morales J, Betanzos A, García-Rivera G, Chávez-Munguía B, Shibayama M, Orozco E. Resveratrol induces apoptosis-like death and prevents in vitro and in vivo virulence of *Entamoeba histolytica*. PLoS One 2016;11(1):e0146287.

[123] Zhang T, Stanley SL. Oral immunization with an attenuated vaccine strain of *Salmonella typhimurium* expressing the serine-rich *Entamoeba histolytica* protein induces an antiamebic immune response and protects gerbils from amebic liver abscess. Infection and Immunity 1996;64(5):1526−31.

[124] Zhang T, Stanley SL. DNA vaccination with the serine rich *Entamoeba histolytica* protein (SREHP) prevents amebic liver abscess in rodent models of disease. Vaccine 1999;18(9−10):868−74.

[125] Carrero JC, Contreras-Rojas A, Sánchez-Hernández B, Petrosyan P, Bobes RJ, Ortiz-Ortiz L, et al. Protection against murine intestinal amoebiasis induced by oral immunization with the 29 kDa antigen of *Entamoeba histolytica* and cholera toxin. Experimental Parasitology 2010;126(3):359−65.

[126] Martínez MB, Rodríguez MA, García-Rivera G, Sánchez T, Hernández-Pando R, Aguilar D, et al. A pcDNA-Ehcpadh vaccine against *Entamoeba histolytica* elicits a protective Th1-like response in hamster liver. Vaccine 2009;27(31):4176−86.

[127] Kaur U, Khurana S, Saikia UN, Dubey ML. Immunogenicity and protective efficacy of heparan sulphate binding proteins of *Entamoeba histolytica* in a guinea pig model of intestinal amoebiasis. Experimental Parasitology 2013;135(3):486−96.

Leishmaniasis

2

Nilakshi Samaranayake

Department of Parasitology, Faculty of Medicine, University of Colombo, Colombo, Sri Lanka

2.1 Introduction

Leishmaniasis is a vector-borne disease caused by a unicellular protozoan parasite. Classified as a neglected tropical disease, over one billion people living in endemic areas are at risk of this infection. More than 20 species of *Leishmania* (order Kinetoplastida, family Trypanosomatidae) cause leishmaniasis in humans. The parasite is transmitted by the bite of an infected female phlebotomine sandfly during a blood meal.

Broadly called as leishmaniases, this disease consists of a spectrum of clinical entities and presentations. These multiple disease phenotypes range from cutaneous leishmaniasis (CL) in its simplest form to mucocutaneous leishmaniasis (MCL) with more extensive involvement of the mucosa to visceral leishmaniasis (VL) that would be fatal unless treated early. CL itself has a range of presentations from self-healing lesions to diffuse forms that are resistant to treatment. Even though not associated with mortality, CL receives focus as the commonest form of the leishmaniases and as a cause of stigma and disfigurement. MCL is a more severe and chronic infection where a primary skin lesion is followed by mucosal involvement, commonly the upper airways, after a variable incubation period. In some patients, the initial lesion may be on the mucosa without evidence of a preceding lesion of the skin. VL denotes the dissemination of the parasite with the primary involvement of organs of the reticuloendothelial system and is considered the third most common parasitic cause of death. Its sequelae, post-kala-azar dermal leishmaniasis (PKDL) is a disfiguring cutaneous condition implicated in the transmission of VL. There is equally wide variability in parasite and sandfly species, mostly differing by geographical location, which adds to the complexity of these diseases. The immune responses generated during infection are influenced by these host—parasite vector-derived factors to determine the clinical pathology and outcome of the disease.

The control measures against leishmaniasis have mostly centered around vector control, with early diagnosis and treatment also playing a vital role. Overall, these control measures have been largely successful in some regions such as South Asia while only limited progress has been made in others such as the African region. Only a limited number of drugs are available as treatment options for leishmaniasis, which is further limited by associated drug resistance and side effects. Thus, efforts

Molecular Advancements in Tropical Diseases Drug Discovery. https://doi.org/10.1016/B978-0-12-821202-8.00002-5

are being made to develop newer drugs as well as better diagnostic tools to make drug treatment more targeted and cost-effective. Evidence of naturally developing immunity and limited antigenic variability of parasite forms in the human host makes the development of a successful vaccine a possibility and would be a worthwhile prophylactic measure, especially in settings of limited success in control activities.

2.2 Dotootion of infection

A patient with leishmaniasis would typically be resident in an endemic area or have visited such an area. CL typically presents with a nonitchy, nontender papule that enlarges slowly over weeks to months with central ulceration. MCL presents as ulcers of the mucous membranes of the nose, mouth, or pharynx, which progress to destroy the soft tissue in the nasal and oral regions. VL may present as a more prolonged illness with intermittent bouts of fever, malaise and loss of weight and hepato-splenomegaly on examination. Although the clinical presentation per se would vary with the type of leishmaniasis, which thus would also influence the differential diagnosis, living in close proximity to an already diagnosed patient with leishmaniasis, immunosuppression and many other risk factors as outlined in Chapter 3 would also alert the clinician to the likely diagnosis.

2.2.1 Laboratory diagnosis

The diagnosis of leishmaniasis is confirmed by visualizing and/or isolating the parasite. Molecular techniques are now commonly used supplementary tests. In CL and MCL, the needle aspirates, slit skin smears, dermal scrapings, and punch biopsies are the primary sample types used for these procedures. Aspirates and biopsies from organs of the reticuloendothelial system such as spleen, liver, and bone marrow are the samples of choice in VL while peripheral blood may also be used but with much lower success rates.

2.2.1.1 Demonstration and isolation of the parasite

The smears prepared with the above types of samples are stained with Leishman, Giemsa, or Wright stains and examined under oil immersion microscope for the amastigote stage of the parasite. Impression smears, made by a slide directly placed over an open lesion or using fresh biopsy specimens, could also be observed for parasites in the same manner. Amastigotes are typically seen within monocytes and macrophages as round or oval bodies (LD bodies); 2—4 μm in diameter with indistinct cytoplasm, a nucleus, and a small, rod-shaped kinetoplast at right angles to the nucleus. Less frequently, amastigotes may be seen lying free between cells.

The first description of *Leishmania* parasites was in histological tissues from a lesion identified as Delhi boil. The amastigote form of the parasite was described in the epidermal and dermal tissue of these lesions by David Cunningham, a

Professor of Medicine and Pathology practicing in India in 1885 [1]. A spectrum of changes has been described in histological specimens taken from cutaneous lesions. Cutaneous lesions of leishmaniasis are characterized by a chronic inflammatory reaction with the presence of lymphocytes, plasma cells, and macrophages. Epithelioid cells and the formation of granulomas, frequently associated with giant cells, hyperkeratosis, follicular plugging, vasculitis, and necrosis, are some of the many observed histological findings in the epidermis and dermis. Histological findings in CL have been compared to that of the spectrum seen in leprosy ranging from lepromatous to tuberculoid leprosy [2,3].

Needle aspirates or tissue homogenates can be inoculated into a variety of artificial culture media such as RPMI 1640, Schneider, M199, modified Novy, Nichole, and MacNeal with 10%−20% heat-inactivated fetal calf serum supplements at 22−26°C for the isolation of promastigotes. The addition of human urine has also been successfully tried out by many laboratories for optimizing growth. The cultures are observed regularly under an inverted microscope or by the wet mount of culture fluid for parasite growth for about 2−4 weeks (based on local parasite growth patterns) before determining as negative. Microculture and miniculture methods have been tried out as a more economical method and with less likelihood of bacterial and fungal contamination [4,5]. Although not routinely practiced for diagnostic purposes, the in vivo culture of the parasite can also be carried out after the inoculation of laboratory animals (e.g., hamsters, mice, or guinea pigs) with infected specimens by different routes such as subcutaneous, intraperitoneal, or intracardiac injection.

2.2.1.2 Immunologic diagnostic methods

Antibody tests targeting the humoral immune response in VL have been widely used in different settings and are based on four major formats: direct agglutination tests (DAT), indirect immunofluorescence, ELISA, and immunochromatography. In DAT, the antigen is prepared from formalin-killed promastigote stages of parasite cultures and stained blue for visibility. The test is semiquantitative. Although the test has high sensitivity and specificity, the use is limited by requirements of laboratory facilities, cold chain maintenance, and trained staff. DAT using freeze-dried antigen as opposed to a liquid antigen is preferred under field conditions. ELISA tests have also been widely tried in the diagnosis of leishmaniasis. This technique is highly sensitive but its specificity depends upon the antigen used, and thus tests based on recombinant antigens as opposed to crude soluble antigen prepared from whole promastigote lysates are generally preferred. A recombinant antigen, rK39, based ELISA is 100% sensitive and 100% specific in the diagnosis of VL caused by parasites of *L. donovani* complex and is also of prognostic value as titers correlate well with disease activity. However, the above is mostly applicable to South Asia and several other antigens have been used to optimize ELISA-based testing in other endemic settings [6−8]. Most of these ELISA tests can be carried out on plasma, serum, or blood spots collected on filter paper as well as urine.

An immunochromatographic strip test based on rK39 antigen is now widely used as a cost-effective, reliable, and easy to perform the point-of-care test in the diagnosis of VL and is considered a first-line test to be done in a clinically suspected VL patient, who has not been previously treated. The procedure of the rK39 dipstick is simple with results available within 20−25min. Although very high sensitivity and specificity are reported for this test in India [9], much lower sensitivities have been reported in Sudan, another focus with *L. donovani* parasites, similar to ELISA tests based on this antigen [10]. In general, all the above antibody tests have the inherent limitation of reliably differentiating present from past infections.

Antigen detection is preferred over antibody-based immunodiagnostic tests due to specificity, but these tests are less widely available for the diagnosis of leishmaniasis. A latex agglutination test to detect urinary antigen during the acute stage of VL has shown good agreement with microscopy and is considered a simple addition to the diagnostics of VL at field level [11,12]. More recently, a rapid diagnostic test based on antigen detection has been developed for the diagnosis of cutaneous leishmaniasis. This immunochromatographic strip test targets a membrane-based amastigote antigen (peroxidoxin) of *Leishmania* species; although it is claimed to be of high sensitivity in detecting infections by *L. major*, it has shown variable performance in other settings [13,14].

2.2.1.3 Skin testing

The Leishmanin skin test (Montenegro test) based on a delayed-type hypersensitivity test, similar to the tuberculin test, has been used in different settings in epidemiological studies as an indicator of exposure. This test will be negative in active VL and becomes positive as the patient recovers. The Leishmanin test is not advocated any longer due to constraints in standardizing the antigen as well as due to safety concerns.

2.2.1.4 Nucleic acid amplification tests

PCR-based methods targeting different genomic regions have been developed by many groups for the diagnosis of leishmaniasis and have been applied to a wide variety of clinical materials [15]. The most commonly used amplification targets are nuclear DNA such as the small subunit ribosomal RNA gene, extrachromosomal DNA such as repetitive kinetoplastid DNA (kDNA), mini-exon genes, and the ribosomal internal transcribed spacer region [16]. Assays that amplify conserved sequences found in minicircles of kinetoplast DNA (kDNA) of *Leishmania* have proven to be the most sensitive in many applications and have also been used to differentiate the parasite at the species level [17]. The higher sensitivity is attributed to the large number of copies of minicircle DNA estimated to range from 10,000 to 20,000 copies/parasite. PCR-based techniques become vital diagnostic tools in situations of low parasitemia such as in subclinical infections and in VL-HIV coinfections where the clinical presentations and immunodiagnostics may be unreliable [16].

The use of PCR-based diagnostic tests is limited by the availability of laboratory resources and trained technical staff as well as being time consuming. More recently, these have been adapted for point-of-care testing in resource-limited settings using isothermal platforms such as loop-mediated isothermal amplification (LAMP) method [18,19]. The LAMP techniques have now been further optimized for field settings using crude DNA extracts [20] and have been used in combination with quick DNA extraction approaches to comprise mobile suitcase laboratories [21,22].

2.3 Epidemiology and risk factors

The leishmaniases are endemic in 98 countries spread over and three territories on five continents (Africa, Asia, Europe, North America, and South America) with a total of over 350 million people at risk [23]. An estimated 700,000 to 1 million new cases and some 26,000−65,000 deaths occur annually due to leishmaniasis while the disability-adjusted life years (DALYs) lost due to this disease is estimated to be close to 2.4 million. Although improved global surveillance has contributed to more reliable data regarding the global status of leishmaniasis the reported numbers are still considered to be underestimated.

2.3.1 Clinical presentations and geographical distribution

The distribution and disease burden of leishmaniases are often classified by clinical phenotype as well as the WHO regions. Visceral leishmaniasis, the most serious form of the disease, is estimated to be responsible for 50,000−90,000 new cases per each year. In 2017, more than 95% of new cases reported to WHO occurred in 10 countries: Bangladesh, Brazil, China, Ethiopia, India, Kenya, Nepal, Somalia, South Sudan, and Sudan.

Cutaneous leishmaniasis, the most common form of the disease, is reported from Central Asia, Middle East, and Mediterranean countries as well as the Americas with six countries: Afghanistan, Algeria, Brazil, Colombia, Iran, Iraq, and the Syrian Arab Republic reporting over 95% of the global burden. MCL is the most limited in numbers of the three clinical presentations. Over 90% of mucocutaneous leishmaniasis cases occur in Bolivia, Brazil, Ethiopia, and Peru [23]. PKDL, a complication of visceral leishmaniasis and now considered a fourth clinical form of leishmaniasis, is commonest in Sudan in East Africa, and in Bangladesh on the Indian subcontinent [24,25].

In any endemic focus, only a small proportion of *Leishmania* infections may manifest as overt disease. Reports of the ratio of asymptomatic to clinical disease have varied from 4:1 [26] to as much as 50:1 [27]. Age patterns in asymptomatic infection prevalence reported from the Indian subcontinent, while not definitive, have suggested increased acquired immunity with age [28]. Such persons with asymptomatic infections are considered to act as reservoirs of infection and contribute to sustaining the transmission cycle even though the efficacy of transmission in such instances is still not well established.

In a backdrop of an increasing population of immune-compromised persons, especially those infected with HIV, the visceral form of the disease has emerged as an important opportunistic infection. Since the end of the 1980s, as many as 35 countries throughout the world have reported cases of VL/HIV coinfection, with the main focus being countries in Southern Europe, African countries such as Ethiopia and Sudan, India in South Asia, and Brazil in Latin America [29]. A concomitant HIV infection is considered to increase the risk of an asymptomatic infection progressing to active VL by 100-fold or more while the relatively high parasite loads in these patients make them ready reservoirs of infection.

2.3.2 Parasite

Over 20 species of Leishmania species have been identified to cause infection in humans. These include the *L. donovani* complex with two species (*L. donovani, L. infantum* [also known as *L. chagasi* in the New World]), the *L. mexicana* complex with three main species (*L. mexicana, L. amazonensis,* and *L. venezuelensis*), *L. tropica, L. major, L. aethiopica* and the subgenus *Viannia* with four main species (*L. [V.] braziliensis, L. [V.] guyanensis, L. [V.] panamensis,* and *L. [V.] peruviana*) [30].

Old World CL is caused mainly by *L. major* and *L. tropica* and less commonly by *L. infantum* and *L. aethiopica*. The main aetiologic species in the New World (the Americas) belong to either the *L. mexicana* species complex or *L. (V.) braziliensis* species complex. Causative parasites of New World CL consists mainly of *L.L. mexicana, L.L. amazonensis,* and *Viannia* subgenus. Mucocutaneous leishmaniasis traditionally refers to a metastatic sequel of New World cutaneous infection. It is caused by species in the Viannia subgenus (especially *L. [V.] braziliensis* but also *L. [V.] panamensis* and *L. [V.] guyanensis*). Visceral leishmaniasis usually is caused by *L. donovani* and *L. infantum* while PKDL is best described in cases following *L. donovani* infection.

Although typically these parasite species are classified in association with clinical phenotypes as mentioned above, more recent approaches based on sequence-based typing of genes to genomes have generated information on greater genetic diversity and phylogenetic relationships among these species.

2.3.3 Vector

The only known vector of the leishmaniases is the small dipteran fly known commonly as a "sandfly." Although there are over 600 species of sandflies only a few dozen species are known to be vectors of human disease [31]. Leishmaniasis is transmitted by various species in the genus Phlebotomus in the Old World and by Lutzomyia species in the New World. Each sandfly species typically transmits only one species of the parasite [32].

Sandflies breed in warm moist environments rich in organic matter. Different from mosquitoes, the immature stages do not require standing water to complete development. Animal burrows, termite mounds, and tree holes are some common breeding sites while debris rich in decaying organic matter, contaminated soil of domesticated animal shelters, earthen floors, or unplastered walls of human habitations have also been implicated [33]. Sandflies are weak flyers that usually fly close to the ground in short hops and generally have a flight range of about 300 m. Peak biting activity of sandflies is from dusk to dawn.

2.3.4 Host/reservoir hosts

Reservoir hosts are important for maintaining the life cycle of *Leishmania* parasites. Depending on the geographical location and the type of transmission, a large number of animals ranging from rodents, mongoose, foxes, jackals, wolves, bats, and primates in more rural/sylvatic infections to dogs, cats, goats, and cows in more peri-domestic cycles have been considered as reservoir hosts [34,35]. The reservoir of infection for Indian kala-azar is humans, while patients with PKDL are also considered a reservoir for transmission of VL [36].

2.3.5 Life cycle and transmission

Female sand flies acquire the amastigote form of *Leishmania* parasites when they feed on an infected mammalian host (Fig. 2.1). Uptake of the parasites, which would

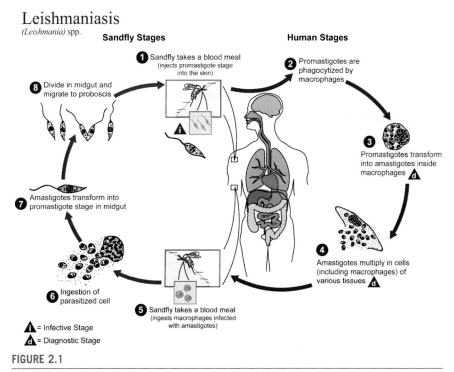

FIGURE 2.1

Life cycle of *Leishmania* parasites.

Source: (https://phil.cdc.gov//PHIL_Images/3400/3400_lores.jpg; Image credit – Alexander J. da Silva, PhD.)

usually be in the macrophages of the skin, by a blood-feeding sandfly is assisted by the cutting action of its mouthparts [32]. In the sandfly's midgut, the parasites transform into promastigotes and undergo further differentiation. Metacyclic promastigotes, the final stage of this differentiation that is infective to mammals, move to the foregut of the sandfly and are deposited in the skin of a new host when the fly takes another blood meal, leading to the transmission of disease [37]. Uncommon modes of transmission include blood transfusion or organ/tissue transplantation [38], contaminated needle sticks, via accidental laboratory exposure [39], congenital transmission and sexual intercourse [40].

2.3.6 Risk factors

The ecology and epidemiology of leishmaniases are affected by interactions between the parasite, host, vector, and the environment. Being a resident in an endemic region or frequent visits or long stays in such an area is foremost among the risk factors for acquiring disease that also suggests the likely repeated exposure to the vector. Generally, environmental conditions and human behavior are among those factors that pose major risks for leishmaniasis in humans across different foci [41,42]. For instance, in a rural setting of transmission, occupations with prolonged outdoor exposure such as farming would increase vector—host contact whereas deforestation would lead to alterations in the ecological niche of the vectors as well as the availability of natural hosts for feeding, with the same end result. Similarly, in a more urban setting of transmission, crowded living conditions with sandfly breeding sites in proximity, which may be due to unplanned urbanization, human migration, or shelters for those displaced by war, are common man-made risk factors. Moreover, if the transmission is anthroponotic, this would add to the risk of transmission of disease in such a setting. Low socioeconomic status is a recognized risk factor for leishmaniasis, which is a likely result of a combination of factors such as poor living conditions conducive for sandfly breeding and lowered immunity owing to poor nutritional status.

Climate change can affect the epidemiology of leishmaniasis in several ways. Changes in rainfall, temperature, and humidity can affect breeding cycles of the vector and can also affect the distribution of both vector and the reservoir host. An increase in the reported number of cases following the rainy seasons has been a common observation in some endemic areas, which is a likely reflection of increasing breeding sites for the vector. Drought, famine and floods can lead to the mass migration of people to areas with active transmission of leishmaniasis resulting in multiple risk factors for the affected population.

In certain groups, the risk factors may be associated with the route of transmission. In Spain, a country that reported a large number of VL-HIV coinfection, a majority of infections is reported to occur in intravenous drug users where shared needles are implicated in the person-to-person transmission of leishmaniasis [43,44]. Economic migration of people predisposed to HIV infection, where they become infected with HIV in urban areas and then return to rural areas where VL is prevalent, has also been reported to increase VL-HIV coinfections [45]. Overall, demographic features in many foci indicate young adult males to be the most at risk of leishmaniasis, which is a likely result of outdoor exposure due to occupation and lifestyle. The increased risk associated with having another affected family member is usually attributed to the exposure to the same sandfly breeding habitats. The susceptibility of some populations to leishmaniasis has also been linked to host genetic factors.

Although acquiring the disease is influenced by these generic risk factors, development into different clinical forms is associated with multiple determinants. *Leishmania* possess a variety of virulence mechanisms that enable the amastigote stage to survive in the host macrophage that is also the key effector cell in killing this parasite. Sandfly saliva can also act as a virulence factor by enhancing the invasion of macrophages by promastigotes. In animal models of immunity against leishmaniasis, commonly studied using *L. major* causing cutaneous lesions, susceptibility to infection is associated with a Th2 type immune response and resistance is related to a Th1 type response. However, this clear-cut dichotomy is often not observed in human infections, suggesting a more complex immune response.

2.3.7 Spatial clustering, outbreaks, and epidemics

In the geographical and regional variables, the main factor for the distribution of leishmaniasis is the presence of vector sandfly in the specific areas. Clustering of cases that reflects the suitable microhabitats of sandfly is a well-recognized feature in the observed geographical spread. Large outbreaks or epidemics of leishmaniasis due to VL or CL have been usually due to mass population movements, for instance, due to war and civil unrest in Sudan and Afghanistan. Large numbers of cases have also been reported in military personnel serving in endemic regions [46,47].

2.4 Approaches to control and elimination

Leishmaniasis continues to affect millions of mostly the poor in Africa, Asia, and Latin America. Activities undertaken for control and elimination received a major boost with a resolution adopted at the Sixtieth World Health Assembly in 2007, which suggested a framework of activities for countries where leishmaniasis was endemic. WHO was also brought to the forefront in supporting these activities by being given the mandate to establish effective control programs in affected countries and to provide relevant technical assistance. These initiatives have been developed further since then with the WHO Roadmap on neglected tropical diseases (2012–20). The existing guidelines are often highlighted by different geographic regions, which is also an indicator of the complex interactions of parasite–host-vector in different foci. The control and elimination strategies adopted in these different settings can be considered under several broad categories.

2.4.1 Early diagnosis and treatment

Early diagnosis plays an important role in control strategies and applies to all clinical types: in VL and MCL to reduce mortality, in CL to reduce morbidity, and in PKDL to reduce the availability of reservoir hosts. Diagnosis should be followed by prompt treatment and monitoring for cure and possible side effects to complete effective case management. An agreed case definition for the country or locality would lead to the early identification of patients even in a resource-poor setting. For instance, in India, the history of fever of more than 2 weeks and splenomegaly in a patient from an endemic area would prompt the healthcare worker to start with a presumptive diagnosis of VL. This approach to a clinical diagnosis must be complemented by other diagnostic tests adapted for point of care testing, for example rK39 immunochromatographic test in India and DAT in Sudan. Availability and better access to medicine for the patients have been improved through WHO mediated programs. Drugs that do not require cold chain preservation and that therefore can be administered even at peripheral centers add to the efficacy of these approaches.

2.4.2 Effective disease surveillance

Disease surveillance includes both active and passive approaches. In most countries, surveillance through passive case detection is done via state-run hospitals and clinics that in turn will be linked to a central data collation system. Regulations to make notification of this disease, mandatory by health practitioners, further improve this process. Although the true burden of disease may not be captured due to reasons such as under-reporting and those who may seek other modalities of treatment, passive surveillance provides the main source of data to monitor disease trends in a locality.

Active case detection is often purpose-driven as in part of epidemiological studies undertaken by research groups and institutions or may be initiated by health authorities itself in instances such as suspected outbreaks. Active case detection becomes more important as the number of cases in a country or region declines and as goals change from control to elimination. Early identification of changing disease-related patterns such as a poor response to treatment or an increase of unusual clinical presentation that may, in turn, indicate changes in parasite and host characteristics is dependent on the regular evaluation of surveillance data. Online epidemiological surveillance systems such as District Health Information System 2 can vastly improve and complement the existing systems by enabling the prompt collection, flow, analysis, and sharing of data by national programs.

2.4.3 Vector control

Vector control and surveillance become the single most important preventive strategy in controlling leishmaniasis, especially in domestic and peridomestic transmission habitats. Integrated vector control that includes a multipronged approach such as indoor residual spraying, destruction of vector breeding sites, and reducing human/vector contact through methods such as insecticide-treated nets (ITNs) is advocated.

Sandfly control depends on its behavior in a particular locality and requires a knowledge of the species, transmission cycles, habitats, host feeding preferences, resting sites, and seasonality for the design of strategies most effective for the focus [48]. Different classes of insecticides are used for indoor residual spraying in different settings. The insecticides used in a given locality should be rotated to prevent the development of resistance and the spraying activities should be complemented with GIS mapping for more effective results. Dog collars impregnated with pyrethroid insecticides have been used for the control of both CL and VL in settings of canine parasite reservoirs. Overall, still, there is a dearth of information on insecticide resistance patterns of sandflies. The success of ITNs would also depend on the successful change in sleeping behavior and acceptability.

2.4.4 Control of reservoir hosts

There is no single strategy for this control approach, and the activities undertaken depend on the local disease dynamics. The focus on reservoir hosts in controlling leishmaniasis has been mainly about zoonotic CL and VL. Domestic dogs in such settings are subject to mass screening by serology as well as a clinical examination by veterinarians for suggestive signs such as loss of fur, wasting, lymph node enlargement and eye discharges. Although culling may be an acceptable method of controlling infected stray dogs, this can often meet with community resistance where domestic dogs are concerned. Attempts have been made to kill some wild animal reservoirs such as gerbils and sand rats by identifying their characteristic burrows and poisoned food baits.

In foci with anthroponotic transmission cycles, all measures related to case management such as early diagnosis and treatment and behavioral modifications to reduce man-vector contact will contribute to effective control measures. PKDL patients are a special target group in control of human reservoirs, as skin lesions in this condition are known to carry high loads of parasites.

2.4.5 Social mobilization and strengthening partnerships

Participation and support of the community become essential in ensuring the long-term sustainability of many of the preventive and control measures. Assessments of knowledge, attitude, and practice should be carried out to find out ways to promote the compatibility of practices, customs, and beliefs among various social groups and minorities, with existing options for leishmaniasis control and prevention [48]. Existing networks of public health workers who may have involved with other disease control campaigns of the country such as malaria and leprosy have been effectively utilized in some settings.

Regional partnerships and global collaborations can contribute in many ways for a country to achieve its control and elimination targets. For countries in the same region that will inevitably face the possibility of importing the disease from a neighboring endemic country, it becomes rational and cost-effective to target common goals. India, Nepal, and Bangladesh, the three countries in South Asia, which bear a large component of the global VL burden set up a common agenda to eliminate VL as a public health problem by 2015 by reducing the annual VL incidence to below 1/10,000 people at the upazilla level in Bangladesh, the subdistricts (block public health center) level in India, and the district level in Nepal. Even though achieving this overall goal has been now revised to 2020, many successes have already been achieved with Nepal reaching and maintaining the target for the last 3 years, Bangladesh achieving its target in over 90% of its endemic subdistricts, and India in 70% of Blocks [49].

2.5 Next-generation diagnostics, drugs, and vaccines

2.5.1 Diagnostic needs in leishmaniasis

As preventive and treatment strategies are stepped up, the accurate diagnosis of leishmaniasis that would inform many of these decisions also becomes paramount. Although any new diagnostic test that is developed should be robust, it should also be relatively inexpensive, require minimal operator training, and should be easy to perform under field conditions. As opposed to rapid diagnostic tests for VL that have been in use, similar tests for CL based on antigen detection have now been developed but have shown variability in sensitivity and specificity with parasite species and geographical location.

Improved diagnostic tests are still required to identify asymptomatic infections where the parasite load is much lower while reliable tests to predict clinical prognosis and to monitor response are also a necessity. Synthetic or recombinant antigens based on genome level information that is now available for many species and host immune response and metabolomic profile-based biomarkers are some candidates being explored for such tests.

2.5.2 New approaches in treatment using existing antileishmanials

WHO lists leishmaniasis as one of the neglected tropical diseases for which the development of new treatments is a priority. Despite the comparative advances in diagnostics over time, there are only a few treatment options for leishmaniasis; namely amphotericin and its formulations, pentavalent antimonials, miltefosine, pentamidine, and paromomycin. These drugs are far from optimal due to toxicity, drug resistance and cost thus creating a need for the development of better drugs.

Several clinical trials to try these existing drugs in combination to reduce toxicity and side effects as well as prevent the development of resistance to monotherapy are now in progress. PKDL and VL-HIV have received focus in these trials as clinical entities which still do not have very effective management strategies. Miltefosine has been targeted in combination with other treatment modalities in many of these studies for different clinical forms due to ease of administration as an oral drug. Another treatment option now being widely tried for CL with or without the combination of another treatment is thermotherapy where localized heat (one session, 50°C for 30s) is applied to the lesion using a simple device [50,51].

2.5.3 Drug candidates in development

Several new chemically diverse drugs, targeting different molecular mechanisms of the parasite are reported to be in late stages of development and may be available for use for visceral leishmaniasis within the next decade. The mode of action of some of these drug candidates includes inhibition of *Leishmania* proteasome, inhibition of *Leishmania* CRK12 kinase enzyme, and bioactivation by parasitic nitroreductase NTR2 resulting in reactive intermediates (Table 2.1). Some of these mechanisms continue to be targeted in the further development of new molecules as described in the subsequent sections.

In the drug groups above, Nitroimidazoles show activity against a broad spectrum of microbes including protozoa. The antileishmanial effect of active compounds of aminopyrazoles was discovered through high-throughput screening and hit optimization of a small molecule library [54]. In addition to the above, several other backup candidates of this group are also under study. Lead compounds of chemical classes of nitroimidazoles (DNDI-0690), aminopyrazoles (DNDI-1047), and benzoxaborole (DNDI-6148) have also shown potential as new therapeutic candidates for cutaneous leishmaniasis following in vitro and in vivo studies in murine models [55].

Table 2.1 Preclinical and clinical drug candidates in development for leishmaniasis.

Drug	Class	Mode of action	Target clinical presentation
DNDi-0690	Nitroimidazole	Bioactivation by parasitic nitroreductase NTR2	CL/VL
DNDi-6148	Oxaborole	Unknown but active against leishmania strains	CL/VL
XE408	Proteasome inhibitor	Inhibition of parasitic proteasome	VL
GSK3494245/ 1305143	Proteasome inhibitor	Inhibition of parasitic proteasome	VL
GSK-3186899/ DDD853651	Pyrazolopyrimidine	Inhibition of leishmania CRK12 kinase	VL
DNDi-5561	Aminopyrazole	Unknown	VL
DNDI-1047	Aminopyrazole	Unknown	CL

Adapted from Rijal S, Sundar S, Mondal D, Das P, Alvar J, Boelaert M. Eliminating visceral leishmaniasis in South Asia: the road ahead. British Medical Journal 2019;364:k5224-k5224 and RedeLEISHNetwork, DNDi Planning activities in cutaneous leishmaniasis. In: InfoLEISH. 3rd ed. Rio de Janeiro, Brazil: DNDi Latin America; 2018.

2.5.4 Molecules and pathways for novel drug targets

2.5.4.1 Parasite targets

Exploring biological pathways/cellular structures that differ between the host and parasite has been a common strategy to initiate the identification of druggable targets. A detailed understanding of molecular mechanisms of drug resistance also contributes to this effort as some of the resistance mechanisms also center on these same molecules and pathways. Some key pathways and molecules essential for parasite survival and virulence that have been targeted in attempts to find viable chemotherapeutic alternatives are summarized (Table 2.2).

Enzymes and other molecules of metabolic pathways

Purine salvage pathway: *Leishmania* parasites cannot synthesize purines de novo and are dependent on the host for these requirements. The parasites obtain purines from their host via salvage synthesis of three different degradation products of purines: inosine-50-monophosphate, guanine, and xanthine. Therefore, purine nucleoside and nucleobase transporters that transport these molecules across the plasma membranes play a crucial role in this process [56]. Although these transporters have also been shown to mediate the uptake of several drugs or experimental drugs they are also structurally and functionally different from host proteins that make them good candidates for selective inhibition. Further, silencing the genes implicated in the regulation of this transporter expression has also been explored as potential approaches while enzymes of this pathway such as Guanosine monophosphate reductase have also been suggested as promising targets [57].

Polyamine pathway: The polyamine pathway is important for parasite replication and to the establishment of infection in the host. Polyamines such as L-arginine, a

Table 2.2 Potential drug targets in *Leishmania* parasites.

Target	Description
Purine salvage pathways and nucleoside transporters	*Leishmania* are purine auxotrophs and dependent on the host to obtain purines by salvage synthesis
Polyamine pathway	Essential for parasite replication, growth, and survival
Glycolytic pathway	Glycolysis is the main source of ATP generation
Sterol biosynthetic pathway	Sterols are vital components of the cell membrane
Antioxidant pathway	Vital in managing oxidative stress
Pathways of carbon metabolism	Energy metabolism depends on the carbon sources available in the host
Protein kinases	Regulators of cell division, signal transduction, and many other vital functions
Cell cycle	Includes enzymes of DNA replication, transcription, recombination, and repair
Proteinases	Virulence factors involved in host cell–parasite interactions

precursor of ornithine, putrescine, and spermidine are essential for parasite functions of growth and survival. Thus, this pathway can be targeted at several points such as inhibition of amino acid permease 3 and drugs acting upon enzymes of the polyamine synthesis pathway such as ornithine decarboxylase inhibitors [58]. Furthermore, the recent description of specific arginine sensors in the parasite that helps its survival in host and vector adds to this repertoire.

Glycolytic pathway: In *Leishmania*, glycolysis, another critical metabolic pathway also occurs in the compartmentalized organelles known as glycosomes. This coupled with structural differences from host enzymes and the availability of 3D structures of some of these enzymes have led to attempts to design drugs that would selectively inhibit them [59].

Sterol biosynthetic pathway: Sterols constitute important components of the cell membrane. The absence of major sterols of the parasite, such as ergosterol, from human cells has made this another attractive target. Several classes of drugs that inhibit the enzymes of this pathway have already been explored. Some of these inhibitors have also been observed to act synergistically when used in combination with azoles [60].

Antioxidant pathway: Trypanothione-based thiol metabolism is unique to trypanosomatidae. Trypanothione synthetase (TS) catalyzes the conjugation of spermidine with two glutathione units to produce trypanothione, and it is then maintained in its reduced form by the enzyme trypanothione reductase (TR). This pathway with its key enzymes is crucial for regulating oxidative stress in parasites. The differences in substrate specificity with its homolog in human cells, glutathione reductase, have been used as the basis of drug discovery focusing on selective inhibition of TR [61]. TS is also considered a target for inhibition with several such compounds already studied. Simultaneous inhibition of TS and TR has also been suggested as an attractive approach.

Carbon metabolism: Several pathways of carbon metabolism are important in the proliferation and virulence of the amastigote stage of *Leishmania*. Recent studies have suggested that survival of intramacrophage stages is linked to controlled metabolic processes with a marked decrease in parasite growth rate and carbon metabolism, which makes this pathway another potential target [62].

Protein kinases/kinome

The protein kinases in the parasite are involved in several essential biological processes including metabolism, gene expression, cell proliferation, motility, differentiation, and death and also influence virulence. The enzymes act by phosphorylation of proteins that in turn alter the activity of the proteins. Although the whole group of kinases or the kinome of the parasites have received focus as potential targets of antileishmanial compounds [63], cyclin-dependent kinases, which are regulators of the cell cycle and mitogen-activated protein kinases, are some widely studied kinases. The target of rapamycin kinases has also been targeted due to its virulence determining properties.

Cell cycle of the parasites

DNA topoisomerases (Top) with its essential role in DNA replication, transcription, and maintaining the stability of the genome have been considered in therapeutic strategies where inhibition could cause cell death. Several compounds classified as Top type IB and Top type II inhibitors in *Leishmania* parasites have been thus described as potential antileishmanials [64].

Secretory proteins

Proteinases in *Leishmania* are of four main types—cysteine, serine, aspartate, and metalloenzymes. Due to their role in host—parasite interactions and virulence factors, groups such as cysteine and serine proteinases have been investigated as druggable parasite targets. Much more recently, inhibitors of the proteasome, a multi-subunit protein complex of the parasite that regulates cellular protein turnover and degradation, have shown promising efficacy in preclinical studies [65].

2.5.4.2 Host molecules and host—pathogen interactions as drug targets

Leishmania parasites are well-adapted parasites with a repertoire of mechanisms to evade or modulate the host immune response. The variability of these responses observed in different clinical presentations can be also utilized in exploring novel therapeutic targets (Table 2.3).

Toll-like receptors

Pattern recognition receptors of the innate immune system such as Toll-like receptors (TLRs) have been implicated in both susceptibility and protection for leishmaniasis with TLR 2, 4, 7 and 9 being some of the frequently studied [66]. CpG-D35 oligodeoxynucleotide, a D-type CpG TLR9 agonist that acts by activation of innate immune cells to produce a proinflammatory response, has been suggested as a

Table 2.3 Potential drug targets related to host cell or immune response.

Target	Description
Toll-like receptors	TLRs recognize pathogen-associated molecular patterns (PAMPs) and downstream signaling of TLRs activate innate immune responses and also orient the adaptive immune response
Host cell membrane fusion mechanism	Membrane fusion and vesicle trafficking are essential processes in the formation of the mature phagolysosome
Selected parasite ectokinases	Some kinases modulate host cell signaling by directly modifying host proteins to attenuate any parasite responses
Histone modifying enzymes	Modify host cell histones altering gene expression
miRNA	Parasites alter host miRNA expression to create an environment suitable for intracellular survival

combination option with standard chemotherapy to boost the immune response. This has been advocated for trials in complicated forms of CL including PKDL. Similarly, studies of identified subsets of innate immune cells have suggested TLR2 and TLR4 antagonists to be promising candidates to modulate the proinflammatory responses in CL that may otherwise result in excessive tissue damage [67].

Host cell—parasite interface

Leishmania parasites are adapted to survive in the hostile environment of the phagolysosome of the host macrophage. Cellular functions required for this such as phagocytosis involve a process of membrane fusion, mediated by a family of peripheral membrane proteins. *Leishmania* parasites target the membrane fusion machinery by different mechanisms such as GP63-mediated cleavage and lipophosphoglycan-mediated phagosome remodeling. These mechanisms and molecules form ideal targets for the design of compounds that will disrupt the intracellular niche permissive for parasite survival and replication [68].

Host cell signaling pathways

These pathways can be made therapeutic foci by inhibition of certain parasite kinases such as Casein kinase 1 isoform 2; CK1.2, which has been shown to directly phosphorylate host proteins thus attenuating the immune response to the intracellular parasites.

Host epigenome

Leishmania has been shown to affect the host epigenome as part of its survival strategy to evade host responses by mechanisms that control gene expression at transcriptional (DNA methylation and histone modifications) and posttranscriptional (up/downregulation of miRNAs) levels. As such, host epigenetic reprogramming directed at these molecules as well as drugs that directly target parasite-released histone-modifying enzymes has been suggested as antileishmanial therapies that are less likely to develop resistance [69].

Host-directed immunotherapy

Immunotherapy that targets immune and inflammatory pathways of the host is an alternative or adjunct to routine treatment that is now studied in many infectious as well as non-infectious diseases. In leishmaniasis, the immune response is considered the single most important host factor that determines the varied clinical presentations and disease outcome. Cytokines to boost immune responses (e.g., IL12, IFNg), monoclonal antibodies to block certain cytokines (e.g., IL10, IL4), or receptors are some widely tried approaches [70].

Cell-based therapy using key cells involved in immune responses to *Leishmania* infections such as dendritic cells and macrophages has also been shown to be effective in laboratory studies while the use of these different molecules and cells in combination with standard chemotherapy has also been demonstrated to improve treatment responses [71].

2.5.5 Alternative approaches to de novo drug discovery

In the face of the cost and time involved in de novo identification and validation of new chemical entities, several other approaches have also been considered to complement these efforts.

2.5.5.1 Novel compounds based on known antileishmanials

Pentavalent synthetic antimonials such as sodium stibogluconate and meglumine antimonite are widely used in many parts of the world even though other drugs have replaced them as a drug of choice, especially for VL. The known efficacy of these has led to the study of organic compounds based on this metal, for instance organoantimony(V) and organoantimony(III) carboxylates for antileishmanial activity. Further developments of these compounds such as organoantimony(V) ferrocenyl dicarboxylates have shown promising results as potential drug candidates against *Leishmania* [72]. Organometallic compounds based on other metals such as bismuth, tin-, tellurium-, palladium-, rhodium-, iridium-, ruthenium-, and iron have also been tested for antileishmaniacidal activity. Further work is ongoing in understanding the metabolism and processing of these drugs and assessing their efficacy and safety.

2.5.5.2 Repurposed drugs as potential antileishmanials

Repositioning or repurposing of existing drugs as therapeutic alternatives has become another attractive option. Anticancer drugs (kinase inhibitors, estrogen-receptor modulators), antimicrobials, antihistamines, drugs acting on the central nervous system, antiarrhythmics, and calcium channel blockers are some of the drugs that have been shown to have antileishmanial activity with varying degrees of efficacy [73].

2.5.5.3 Natural products as antileishmanials

Exploration of natural products from different sources such as plants and marine organisms has been presented as another therapeutic option in recent times. A wide range of alternative natural compounds such as alkaloids, chalcones, triterpenoids, naphthoquinones, quinones, terpenoids, sterols, lignans, saponins, and flavonoids have been considered as candidates for screening and further study for the treatment of leishmaniasis. Laboratory studies have also shown the capability of some of these products to reverse multidrug resistance [74].

2.5.6 Factors influencing success in drug development

Being a neglected tropical disease, the therapeutic needs of leishmaniasis has been a focus of large-scale product development partnerships such as Drugs for Neglected Disease Initiative (DNDi). In addition to the many recognized intrinsic and extrinsic factors that affect the successful development of a new drug, progress, and advances in many complementary fields also contribute to these efforts.

Next-generation phenotypic screening approaches where the conventional approach is supplemented by developments in genomics and transcriptomics, availability of a variety of databases with large amounts of biologically relevant information and omics tools that facilitate high throughput screening and automation of analysis pipelines are some such recent developments [75].

Nanodrug delivery systems (NanoDDS) have several characteristics that may be advantageous to newer drugs being developed for leishmaniasis. The use of NanoDDS can improve the solubility of the drug in an aqueous medium. Further, particulate matter uptake mechanism by macrophages can be manipulated to release leishmanicidal drugs in macrophage-rich organs such as liver, spleen, and bone marrow, using this system. Even though different nanosystems have been identified with limitations such as toxicity, safety, and variability of the product, this still has the potential to contribute to the development of drugs with better efficacy [76,77].

Aptamers, short single-stranded RNA or DNA oligonucleotides that bind target molecules with high affinity and specificity, is another novel therapeutic tool with much promise. Although the mechanism of action is similar to monoclonal antibodies, aptamers have several advantages ranging from increased stability to poor immunogenicity due to the nature of nucleic acids. Functional studies conducted using aptamers against poly (A)-binding protein from *Leishmania infantum* (LiPABP) have suggested effects on parasite gene expression in vivo highlighting its potential in drug development [78].

2.5.7 Vaccines for leishmaniasis

The current situation, where control and elimination strategies are progressing well in South Asia while there are challenges in drug delivery and lower success in implementing control measures in some other regions, highlights that there is still a place for the development of a vaccine to reduce or block transmission. As there is variability in parasite species and strains depending on geographical location, the product characteristics should be carefully designed to reflect vaccines that either target common antigens for different species/strains of the parasite or are tailored to fit regional needs [79].

The earliest vaccine attempts in leishmaniases have been against cutaneous disease where material from a cutaneous lesion of a person was deliberately inoculated into the skin in an unexposed area of a healthy person, a practice among communities in endemic regions. Killed whole Leishmania parasites or fractions of the parasites, live attenuated *Leishmania* parasites, recombinant proteins mostly targeting parasite proteins associated with virulence, subunit, virus-like particle, DNA vaccines, and combinations are some candidates that have hitherto been tried in vaccine studies [80−82]. The successful development of vaccines for dogs against leishmaniasis also added to the promising aspect of a VL vaccine. Salivary proteins of sandflies, alone or in a multivalent approach together with *Leishmania*-derived antigens, constitute another group of potential vaccine candidates [83]. More recently, extensively characterized live attenuated *L. donovani* strains have shown promising immunogenicity in animal studies using both naïve as well as preexposed animals with low parasitemia [84].

However, to date, no vaccine for leishmaniasis has been approved for general clinical use. Effective adjuvants and antigen delivery systems are also critical in developing a successful vaccine. Several such antigen delivery systems based on cationic solid lipid nanoparticles and alginate, a linear polysaccharide with immunoadjuvant properties, are candidates under research [81]. Lipid formulation of saponins composed of phospholipids and cholesterol to deliver antigens known as ISCOMs and their matrix formulation (Iscomatrix) have also shown promise as immunogenic lipid-based delivery systems [85].

Improved understanding of the pathogenesis of leishmaniasis and better animal models that more accurately reflect the human disease is a prerequisite to developing a successful vaccine. Knowledge elicited from both parasite and human genomic studies and systems biology approaches coupled with the use of immune-informatic tools to prioritize potential candidates through reverse vaccine approaches is expected to shorten the process of designing and refining future vaccines for leishmaniasis.

2.6 Concluding remarks

Leishmaniasis affects the poorest in the developing world and contributes to considerable morbidity and mortality. The spectrum of clinical presentations that contributes to the variability of this group of diseases is further complicated by the diversity of the parasites, vectors, and the conducive ecological factors in different foci of endemicity.

Although the standard tests of visualizing and isolating the parasite, culture, immunological and molecular methods provide a wide repertoire for conventional diagnosis, markers for identification of asymptomatic infections and for identifying treatment response and relapse have become an urgent need. The control and elimination efforts based on a multipronged approach centering on rapid case detection and vector control have proven very successful in some settings, whereas natural disasters, civil wars, and zoonotic transmission cycles have hampered progress in others.

With the limited number of treatment options and widespread drug resistance, the ideal therapy, especially for VL, would be an efficacious, safe, oral, and short-course combination treatment. To this end, novel chemicals targeting different parasite biological processes as well as novel drug delivery modes are being explored. The above limitations of these drugs including the lack of a chemoprophylactic and the inability to achieve a sterile cure has highlighted the need for a preventive vaccine. A better understanding of pathophysiological and immunological complexities of the varied disease presentations and animal models that closely reflect human immune responses is a prerequisite to developing such a vaccine. The availability of genomes of several *Leishmania* species coupled with developments in the other omics fields has added to the pace of discovery of drug/vaccine targets. Although the strategic approaches and priority interventions may vary with the region, total eradication of leishmaniasis will depend on the combined efforts of governments, scientific researchers, pharmaceutical industry, and the vulnerable community.

References

[1] WHO. The Leishmaniases: timeline of facts. 2018. https://www.who.int/leishmaniasis/disease/Leishmaniasis-interactive-timelines/en/. [Accessed 1 March 2019].

[2] Herath C, Ratnatunga N, Waduge R, Ratnayake P, Ratnatunga C, Ramadasa S. A histopathological study of cutaneous leishmaniasis in Sri Lanka. Ceylon Medical Journal 2010;55(4):106—11.

[3] Venkataram M, Moosa M, Devi L. Histopathological spectrum in cutaneous leishmaniasis: a study in Oman. Indian Journal of Dermatology Venereology and Leprology 2001;67:294—8.

[4] Ihalamulla RL, Rajapaksa US, Karunaweera ND. Microculture for the isolation of Leishmania parasites from cutaneous lesions – Sri Lankan experience. Annals of Tropical Medicine and Parasitology 2005;99(6):571—5.

[5] Boggild AK, Miranda-Verastegui C, Espinosa D, et al. Optimization of microculture and evaluation of miniculture for the isolation of Leishmania parasites from cutaneous lesions in Peru. The American Journal of Tropical Medicine and Hygiene 2008;79(6):847—52.

[6] Zahidul Islam M, Itoh M, Takagi H, et al. Enzyme-linked immunosorbent assay to detect urinary antibody against recombinant rKRP42 antigen made from *Leishmania donovani* for the diagnosis of visceral leishmaniasis. The American Journal of Tropical Medicine and Hygiene 2008;79(4):599—604.

[7] Zijlstra E, Daifalla N, Kager PA, et al. rK39 enzyme-linked immunosorbent assay for diagnosis of *Leishmania Donovani* infection. Clinical and Diagnostic Laboratory Immunology. 1998;5(5):717—20.

[8] Vaish M, Bhatia A, Reed SG, Chakravarty J, Sundar S. Evaluation of rK28 antigen for serodiagnosis of visceral leishmaniasis in India. Clinical Microbiology and Infections 2012;18(1):81—5.

[9] Sundar S, Rai M. Laboratory diagnosis of visceral leishmaniasis. Clinical and Diagnostic Laboratory Immunology 2002;9(5):951—8.

[10] Maia Z, Lírio M, Mistro S, Mendes CMC, Mehta SR, Badaro R. Comparative study of rK39 Leishmania antigen for serodiagnosis of visceral leishmaniasis: systematic review with meta-analysis. PLoS Neglected Tropical Diseases 2012;6(1):e1484.

[11] Attar ZJ, Chance ML, el-Safi S, et al. Latex agglutination test for the detection of urinary antigens in visceral leishmaniasis. Acta Tropica 2001;78(1):11—6.

[12] El-Safi SH, Abdel-Haleem A, Hammad A, et al. Field evaluation of latex agglutination test for detecting urinary antigens in visceral leishmaniasis in Sudan. Eastern Mediterranean Health Journal 2003;9(4):844—55.

[13] Bennis I, Verdonck K, El Khalfaoui N, et al. Accuracy of a rapid diagnostic test based on antigen detection for the diagnosis of cutaneous leishmaniasis in patients with

suggestive skin lesions in Morocco. The American Journal of Tropical Medicine and Hygiene 2018;99(3):716—22.

[14] De Silva G, Somaratne V, Senaratne S, et al. Efficacy of a new rapid diagnostic test kit to diagnose Sri Lankan cutaneous leishmaniasis caused by Leishmania donovani. PLoS One 2017;12(11):e0187024.

[15] Lachaud L, Dercure J, Chabbert E, et al. Optimized PCR using patient blood samples for diagnosis and follow-up of visceral Leishmaniasis, with special reference to AIDS patients. Journal of Clinical Microbiology 2000;38(1):236—40.

[16] Sundar S, Singh OP. Molecular diagnosis of visceral leishmaniasis. Molecular Diagnosis & Therapy 2018;22(4);443—57

[17] Salotra P, Sreenivas G, Pogue GP, et al. Development of a species-specific PCR assay for detection of Leishmania donovani in clinical samples from patients with kala-azar and post-kala-azar dermal leishmaniasis. Journal of Clinical Microbiology 2001; 39(3):849—54.

[18] Sriworarat C, Phumee A, Mungthin M, Leelayoova S, Siriyasatien P. Development of loop-mediated isothermal amplification (LAMP) for simple detection of Leishmania infection. Parasites & Vectors 2015;8:591.

[19] Verma S, Avishek K, Sharma V, Negi NS, Ramesh V, Salotra P. Application of loop-mediated isothermal amplification assay for the sensitive and rapid diagnosis of visceral leishmaniasis and post-kala-azar dermal leishmaniasis. Diagnostic Microbiology and Infectious Disease 2013;75(4):390—5.

[20] Mikita K, Maeda T, Yoshikawa S, Ono T, Miyahira Y, Kawana A. The Direct Boil-LAMP method: a simple and rapid diagnostic method for cutaneous leishmaniasis. Parasitology International 2014;63(6):785—9.

[21] Mondal D, Ghosh P, Khan MAA, et al. Mobile suitcase laboratory for rapid detection of Leishmania donovani using recombinase polymerase amplification assay. Parasites & Vectors 2016;9(1):281.

[22] de Vries HJC, Reedijk SH, Schallig HDFH. Cutaneous leishmaniasis: recent developments in diagnosis and management. American Journal of Clinical Dermatology 2015;16(2):99—109.

[23] WHO. Leishmaniasis: WHO, fact sheet. 2019. https://www.who.int/news-room/fact-sheets/detail/leishmaniasis. [Accessed 5 March 2019].

[24] WHO. Leishmaniasis in high-burden countries: an epidemiological update based on data reported in 2014. Weekly Epidemiological Record 2016;91(22):287—96.

[25] WHO. Post-kala-azar dermal leishmaniasis: a manual for case management and control: report of a WHO consultative meeting, Kolkata, India, 2—3 July 2012. Geneva, Switzerland. 2—3 July 2012.

[26] Schaefer KU, Kurtzhals JA, Gachihi GS, Muller AS, Kager PA. A prospective sero-epidemiological study of visceral leishmaniasis in Baringo District, Rift Valley Province, Kenya. Transactions of the Royal Society of Tropical Medicine and Hygiene 1995;89(5):471—5.

[27] Moral L, Rubio EM, Moya M. A leishmanin skin test survey in the human population of l'Alacanti region (Spain): implications for the epidemiology of Leishmania infantum infection in southern Europe. Transactions of the Royal Society of Tropical Medicine and Hygiene 2002;96(2):129—32.

[28] Chapman LAC, Morgan ALK, Adams ER, Bern C, Medley GF, Hollingsworth TD. Age trends in asymptomatic and symptomatic Leishmania donovani infection in the Indian

subcontinent: a review and analysis of data from diagnostic and epidemiological studies. PLoS Neglected Tropical Diseases 2018;12(12):e0006803.

[29] WHO. Leishmaniasis and HIV coinfection. 2019. https://www.who.int/leishmaniasis/burden/hiv_coinfection/burden_hiv_coinfection/en/. [Accessed 10 March 2019].

[30] CDC. Leishmaniasis biology. 2018. https://www.cdc.gov/parasites/leishmaniasis/biology.html. [Accessed 1 February 2019].

[31] Lewis DJ. Phlebotomid sandflies. Bulletin of the World Health Organization 1971; 44(4):535−51.

[32] Bates PA. Transmission of Leishmania metacyclic promastigotes by phlebotomine sand flies. International Journal for Parasitology 2007;37(10):1097−106.

[33] Feliciangeli MD. Natural breeding places of phlebotomine sandflies. Medical and Veterinary Entomology 2004;18(1):71−80.

[34] Roque ALR, Jansen AM. Wild and synanthropic reservoirs of *Leishmania* species in the Americas. International Journal for Parasitology Parasites and Wildlife 2014;3(3): 251−62.

[35] Bhattarai NR, Van der Auwera G, Rijal S, et al. Domestic animals and epidemiology of visceral leishmaniasis, Nepal. Emerging Infectious Diseases 2010;16(2):231−7.

[36] Desjeux P, Ghosh RS, Dhalaria P, Strub-Wourgaft N, Zijlstra EE. Report of the post kala-azar dermal leishmaniasis (PKDL) consortium meeting, New Delhi, India, 27−29 June 2012. Parasites & Vectors 2013;6:196.

[37] Alemayehu B, Alemayehu M. Leishmaniasis: a review on parasite, vector and reservoir host. Health Science Journal 2017;11(4):519.

[38] Antinori S, Cascio A, Parravicini C, Bianchi R, Corbellino M. Leishmaniasis among organ transplant recipients. The Lancet Infectious Diseases 2008;8(3):191−9.

[39] Herwaldt BL. Laboratory-acquired parasitic infections from accidental exposures. Clinical Microbiology Reviews 2001;14(4):659.

[40] Avila-García M, Mancilla J, Segura-Cervantes E, Galindo-Sevilla NM. Transmission to humans. In: Claborn DM, editor. Leishmaniasis - trends in epidemiology, diagnosis and treatment. IntechOpen; 2014.

[41] Oryan A, Akbari M. Worldwide risk factors in leishmaniasis. Asian Pacific Journal of Tropical Medicine 2016;9(10):925−32.

[42] Desjeux P. The increase in risk factors for leishmaniasis worldwide. Transactions of the Royal Society of Tropical Medicine and Hygiene 2001;95(3):239−43.

[43] Alvar J, Cañavate C, Gutiérrez-Solar B, et al. Leishmania and human immunodeficiency virus coinfection: the first 10 years. Clinical Microbiology Reviews 1997; 10(2):298−319.

[44] Molina R, Gradoni L, Alvar J. HIV and the transmission of *Leishmania*. Annals of Tropical Medicine and Parasitology 2003;97(Suppl. 1):29−45.

[45] Argaw D, Mulugeta A, Herrero M, et al. Risk factors for Visceral Leishmaniasis among residents and migrants in Kafta-Humera, Ethiopia. PLoS Neglected Tropical Diseases 2013;7(11):e2543.

[46] Patino LH, Mendez C, Rodriguez O, et al. Spatial distribution, Leishmania species and clinical traits of cutaneous leishmaniasis cases in the Colombian army. PLoS Neglected Tropical Diseases 2017;11(8):e0005876.

[47] Magill AJ, Grogl M, Gasser Jr RA, Sun W, Oster CN. Visceral infection caused by *Leishmania tropica* in veterans of operation desert storm. New England Journal of Medicine 1993;328(19):1383−7.

[48] WHO. Strategic framework for leishmaniasis control in the WHO European Region 2014–2020. Geneva, Switzerland. 2014.

[49] Sundar S, Singh OP, Chakravarty J. Visceral leishmaniasis elimination targets in India, strategies for preventing resurgence. Expert Review of Anti-infective Therapy 2018; 16(11):805–12.

[50] DNDi. New CL combination therapies. 2019. https://www.dndi.org/diseases-projects/portfolio/new-cl-combos/. [Accessed 10 March 2019].

[51] Gonçalves SVCB, Costa CHN. Treatment of cutaneous leishmaniasis with thermotherapy in Brazil: an efficacy and safety study. Anais Brasileiros de Dermatologia 2018;93: 347–55.

[52] Rijal S, Sundar S, Mondal D, Das P, Alvar J, Boelaert M. Eliminating visceral leishmaniasis in South Asia: the road ahead. British Medical Journal 2019;364:k5224.

[53] RedeLEISHNetwork. DNDi Planning activities in cutaneous leishmaniasis. In: InfoLEISH. 3rd ed. Rio de Janeiro, Brazil: DNDi Latin America; 2018.

[54] Mowbray CE, Braillard S, Speed W, et al. Novel amino-pyrazole ureas with potent in vitro and in vivo antileishmanial activity. Journal of Medicinal Chemistry 2015; 58(24):9615–24.

[55] Van Bocxlaer K, Caridha D, Black C, et al. Novel benzoxaborole, nitroimidazole and aminopyrazoles with activity against experimental cutaneous leishmaniasis. International Journal for Parasitology: Drugs and Drug Resistance 2019;11:129–38.

[56] Ortiz D, Sanchez MA, Koch HP, Larsson HP, Landfear SM. An acid-activated nucleobase transporter from Leishmania major. Journal of Biological Chemistry 2009; 284(24):16164–9.

[57] Boitz JM, Jardim A, Ullman B. GMP reductase and genetic uncoupling of adenylate and guanylate metabolism in *Leishmania donovani* parasites. Molecular and Biochemical Parasitology 2016;208(2):74–83.

[58] Aoki JI, Muxel SM, Fernandes JCR, Floeter-Winter LM. The polyamine pathway as a potential target for leishmaniases chemotherapy. In: Afrin F, Hemeg H, editors. Leishmaniases as Re-emerging diseases. IntechOpen; 2018.

[59] Verlinde CL, Hannaert V, Blonski C, et al. Glycolysis as a target for the design of new anti-trypanosome drugs. Drug Resistance Updates 2001;4(1):50–65.

[60] Chawla B, Madhubala R. Drug targets in Leishmania. Journal of Parasitic Diseases 2010;34(1):1–13.

[61] Kumar S, Ali MR, Bawa S. Mini review on tricyclic compounds as an inhibitor of trypanothione reductase. Journal of Pharmacy & Bioallied Sciences 2014;6(4):222–8.

[62] McConville M, Saunders E, Kloehn J, Dagley M. Leishmania carbon metabolism in the macrophage phagolysosome- feast or famine? [version 1; peer review: 3 approved] F1000Research 2015;4(F1000 Faculty Rev):938.

[63] Borba JVB, Silva AC, Ramos PIP, et al. Unveiling the kinomes of *Leishmania infantum* and L. *Braziliensis* empowers the discovery of new kinase targets and antileishmanial compounds. Computational and Structural Biotechnology Journal 2019;17: 352–61.

[64] Reguera RM, Elmahallawy EK, Garcia-Estrada C, Carbajo-Andres R, Balana-Fouce R. DNA Topoisomerases of Leishmania parasites; druggable targets for drug discovery. Current Medicinal Chemistry 2018.

[65] Wyllie S, Brand S, Thomas M, et al. Preclinical candidate for the treatment of visceral leishmaniasis that acts through proteasome inhibition. Proceedings of the National Academy of Sciences of the United States of America 2019;116(19):9318–23.

[66] Campos MB, Lima LVR, de Lima ACS, et al. Toll-like receptors 2, 4, and 9 expressions over the entire clinical and immunopathological spectrum of American cutaneous leishmaniasis due to *Leishmania* (V.) *braziliensis* and *Leishmania* (L.) *amazonensis*. PLoS One 2018;13(3):e0194383.

[67] Polari LP, Carneiro PP, Macedo M, et al. *Leishmania braziliensis* infection enhances toll-like receptors 2 and 4 expression and triggers TNF-α and IL-10 production in human cutaneous leishmaniasis. Frontiers in Cellular and Infection Microbiology 2019; 9(120).

[68] Descoteaux A. Chapter 19 the macrophage—parasite interface as a chemotherapeutic target in leishmaniasis. In: Rivas L, Gil C, editors. Drug discovery for leishmaniasis. The Royal Society of Chemistry; 2018. p. 387—95.

[69] Afrin F, Khan I, Hemeg HA. Leishmania-host interactions—an epigenetic paradigm. Frontiers in Immunology 2019;10(492).

[70] Dayakar A, Chandrasekaran S, Kuchipudi SV, Kalangi SK. Cytokines: key determinants of resistance or disease progression in visceral leishmaniasis: opportunities for novel diagnostics and immunotherapy. Frontiers in Immunology 2019;10(670).

[71] Taslimi Y, Zahedifard F, Rafati S. Leishmaniasis and various immunotherapeutic approaches. Parasitology 2018;145(4):497—507.

[72] do Prado BR, Islam A, Frézard F, Demicheli C. Organometallic compounds in chemotherapy against leishmania. In: Rivas L, Gil C, editors. Drug discovery for leishmaniasis. The Royal Society of Chemistry; 2018. p. 199—223.

[73] Charlton RL, Rossi-Bergmann B, Denny PW, Steel PG. Repurposing as a strategy for the discovery of new anti-leishmanials: the-state-of-the-art. Parasitology 2018;145(2): 219—36.

[74] Passero LFD, Cruz LA, Santos-Gomes G, Rodrigues E, Laurenti MD, Lago JHG. Conventional versus natural alternative treatments for leishmaniasis: a review. Current Topics in Medicinal Chemistry 2018;18(15):1275—86.

[75] Aulner N, Danckaert A, Ihm J, Shum D, Shorte SL. Next-generation phenotypic screening in early drug discovery for infectious diseases. Trends in Parasitology 2019;35(7):559—70.

[76] de Souza A, Marins DSS, Mathias SL, et al. Promising nanotherapy in treating leishmaniasis. International Journal of Pharmaceutics 2018;547(1—2):421—31.

[77] Makala LHC, Baban B. Novel therapeutic approaches to *Leishmania* infection. In: Claborn DM, editor. Leishmaniasis - trends in epidemiology, diagnosis and treatment. IntechOpen; 2014.

[78] Guerra-Pérez N, Ramos E, García-Hernández M, et al. Molecular and functional characterization of ssDNA aptamers that specifically bind *Leishmania infantum* PABP. PLoS One 2015;10(10):e0140048.

[79] Mo AX, Pesce J, Fenton Hall B. Visceral leishmaniasis control and elimination: is there a role for vaccines in achieving regional and global goals? The American Journal of Tropical Medicine and Hygiene 2016;95(3):514—21.

[80] Srivastava S, Shankar P, Mishra J, Singh S. Possibilities and challenges for developing a successful vaccine for leishmaniasis. Parasites & Vectors 2016;9(1):277.

[81] Ghorbani M, Farhoudi R. Leishmaniasis in humans: drug or vaccine therapy? Drug Design, Development and Therapy 2017;12:25—40.

[82] Cecílio P, Oliveira F, da Silva AC. Vaccines for human leishmaniasis: where do we stand and what is still missing? In: Afrin F, Hemeg H, editors. Leishmaniases as Re-emerging diseases. IntechOpen; 2018.

[83] Coutinho-Abreu IV, Valenzuela JG. Comparative evolution of sand fly salivary protein families and implications for biomarkers of vector exposure and salivary vaccine candidates. Frontiers in Cellular and Infection Microbiology 2018;8(290).

[84] Ismail N, Kaul A, Bhattacharya P, Gannavaram S, Nakhasi HL. Immunization with live attenuated *Leishmania donovani* centrin(-/-) parasites is efficacious in asymptomatic infection. Frontiers in Immunology 2017;8:1788.

[85] Sabur A, Asad M, Ali N. Lipid based delivery and immuno-stimulatory systems: master tools to combat leishmaniasis. Cellular Immunology 2016;309:55−60.

Tuberculosis

3

Anupam Jyoti*, Sanket Kaushik*, Vijay Kumar Srivastava*

Amity Institute of Biotechnology, Amity University Rajasthan, Jaipur, India

3.1 Introduction

Tuberculosis (TB) is caused by a bacterium called *Mycobacterium tuberculosis*. The bacteria primarily attack the lungs but can attack any part of the body such as the kidney, spine, and brain. If not treated properly, TB disease can be fatal. It is one of the world's deadliest contagious diseases and causes severe health concerns worldwide including India because of a significant increase in multiple drug-resistant (MDR) *M. tuberculosis* varieties [1]. According to the World Health Organization [2], approximately one-third of the world's population is infected by the bacteria and South Africa is one of the worst affected country and was the origin for an HIV-associated, extensively drug-resistant TB (XDR-TB) outburst in 2005 within the KwaZulu Natal province [3]. However, MDR and extensively XDR-TB forms of TB continue to emerge making TB much more difficult to treat and threatens TB elimination efforts. According to the WHO 2014 [3], there are 480,000 cases of MDR-TB and almost there were 210,000 deaths reported in 2013. XDR-TB and MDR-TB have been identified throughout the world, and approximately 9% of patients having MDR-TB have XDR-TB that is recognized by resistance to MDR-TB drugs resulting in the poor treatment of the patients [2]. For most of the individuals suffering from TB, first-line therapy begins with the combination of the four drugs that are isoniazid (INH), rifampicin (Rif), ethambutol, and pyrazinamide that is given for 2 months to remove most of the infection as quickly as possible, and this 2-month phase of treatment is known as bactericidal phase [2]. In another 4 months, the patient continues with the two drugs Rif and INH, expected that, in the initial 2 months of the treatment, major infection is cleared [5]. After the X-ray of the chest, if TB persists, the patient goes for the longer treatment that results in the side effects from the medication and length of treatment causing hepatitis, gastrointestinal intolerance, rash, and renal failure [6], which will lead to the MDR-TB. The MDR patient resistant to Rif and INH goes through the second line of treatment consisting of pyrazinamide, a fluoroquinolone, an injectable antibiotic (amikacin or kanamycin or capreomycin), ethionamide (or

---

* Equally contributed.

prothionamide), and either cycloserine or *para*-aminosalicylic acid with treatment span lasting for at least 18–20 months [5].

Most of the drugs against the pathogenic *Mtb* were discovered more than four decades ago but still comprise the foundation therapy, with the exemption of bedaquiline and delamanid that have established as a treatment for MDR-TB recently [7–9]. After a long decade, yet there is no new drug that has been discovered despite the availability of the sequence of the whole *M. tuberculosis* genome [1] and the host *Homo sapiens* [10]. With the availability of the whole genome of host and pathogen, it has now become easier for the scientist and researchers to search for a new pathway that remains unexplored yet and new enzymes that can be potential drug targets, not present in the host [11]. It is a need of the future to understand the mechanism of action in the host and pathogen to overcome the MDR and XDR. Besides the development of several new techniques, high throughput screening of drugs can be performed at the initial stages of drug development. In today's world, there is an urgent need to combat the *Mtb* with the help of new drugs that are not resistant to *Mtb*.

3.2 Detection of infection

TB is a leading cause of death globally after acquired immunodeficiency syndrome despite medical advances, such as the new generation of antibiotics and other chemical therapies. One of the greatest limitations in the management of TB is its rapid and specific diagnosis. The delay in diagnosis invariably leads to the proliferation and spread of *M. tuberculosis* that results in increased patient mortality. Under such circumstances, there is an increasing demand for early detection, and reliable diagnostic methods for TB that can achieve improved results as well as reduce the number of deaths and medical costs. Several tests and diagnosis parameters including radiological screening, lipoarabinomannan (LAM) assay, and bacteriological testing have been set to detect the infection of *M. tuberculosis* (Fig. 3.1).

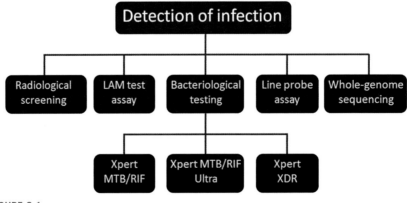

FIGURE 3.1

Figure showing different methods of *Mycobacterium tuberculosis* detection.

All these methods complement each other for specific, sensitive, and rapid diagnosis of tuberculosis.

Radiological screening: With the use of a chest X-ray followed by the computer-aided tool, the detection of pulmonary tuberculosis infection is relatively easy these days [12]. Looking at the success of this method especially for prisoners, mining sector workers, and cementing sector workers, WHO have approved this method of TB detection [13].

LAM test assay: LAM, a cell wall glycolipid secreted by *M. tuberculosis* during infection followed by replication in the host has been used as indicative of TB with renal involvement [14]. This LAM-based assay for the detection of TB is commercially available because LAM is secreted in urine as the bacteria enter the kidney in patients of HIV-associated TB [15]. This LAM-based assay for the detection of TB in HIV patients is rapid, highly economical, and sensitive, linked with lower death rate [16−18].

Bacteriological testing: Several assays/kits encompassing various bacterial factors that affect the growth and pathogenesis of *M. tuberculosis* are used for the detection of *M. tuberculosis* bacterium [19]. The development of Xpert *MTB*/RIF (Cepheid, CA, USA) test has rendered the detection of *M. tuberculosis* along with the genetic mutations conferring rifampicin resistance [20]. However, the major limitation of this test is the longer detection time for its wider application [21]. Keeping this in mind, a newer version of this assay named Xpert *MTB*/RIF Ultra, which is less time consuming and highly sensitive, has been developed; hence, it has been recommended by WHO for the detection of *M. tuberculosis*. Furthermore, Xpert XDR that can detect *M. tuberculosis* and mutations responsible for resistance against isoniazid and fluoroquinolones has been developed [22,23,24]. Moreover, the development of portable kit for the onsite detection of *M. tuberculosis* is required.

Line probe assay: In this assay, the biotinylated-labeled PCR amplicons of a gene responsible for mediating antibiotic resistance are hybridized with probes followed by colorimetric detection. Based on the color band, mutations conferring antibiotic resistance are identified. Line probe assays are used for the detection of rifampicin and isoniazid resistance in *M. tuberculosis*. Earlier, this assay was used for smear-positive specimens [12,13]. Further, with the introduction of an advanced version of this assay, the detection of *M. tuberculosis* in smear-positive and smear-negative samples has become possible [25].

Whole-genome sequencing: With the development of whole-genome sequencing methods, the specific detection of *M. tuberculosis* and mutations associated with drug resistance is easier [26]. This technology has immense potential for the detection of *M. tuberculosis* even in the disease epidemic [27].

3.3 Epidemiology and risk factors

Globally, TB is the foremost cause of death among single infectious agent after HIV [28]. The incidence of mortality caused by TB was 1.4 million in 2016, which continues to rise [29]. Epidemiology of TB is not restricted to a region or specific country, rather it has widespread coverage. The incidence of TB is more prevalent in Africa followed by India, China, Indonesia, Philippines, Pakistan, Nigeria, and Bangladesh [30]. Few cases have been reported in America and European countries. This suggests that *M. tuberculosis* is more active in the tropical and subtropical regions as compared with the temperate region [31].

Paleopathological evidence from 7000 BCE suggested that TB was a very common cause of death, which has reduced significantly over several millennia [32]. The incidence of TB has decreased (nearly 10% decline per year in Europe) during the 20th century especially after the Second World War [33]. On the contrary, the incidence of TB increased dramatically after the Second World War resulting in a large proportion of the population infected with TB. With the increase in the number of HIV-infected patients in 2001, the number of TB incidence has augmented globally. This is because the chances of TB infection are 20 times more in HIV-infected patients compared with people without HIV infection. Since 2006, the number of TB patients has decreased at a rate of 2% annually. For instance, there were nearly 14 million cases of TB that were reported, which further lowered down to nearly 8 million in 2010. Economic development, improved nutrition, better hygiene, rapid diagnosis of *M. tuberculosis*, and development of specific chemotherapy were the major attributes toward the reduction in the number of TB cases worldwide. Similarly, incidences of TB mortality cases have reduced to 45% including HIV-associated cases. Even if the transmission of TB completely stops, a large number of TB cases can be reported due to a very large pool of infected individuals [34].

Demographically, the distribution of TB patients is uneven worldwide. The most prevalent incidence of TB patients is from Asian and African countries that account for nearly 80% of the world's total population. In the past few years, nearly 80% reduction in TB-associated death rate was reported in China. The highest incidence of TB has been reported in Swaziland, the African continent where the national average was 1200 cases per 100,000 populations in 2007. In European countries, the incidence of TB is relatively very low mainly ranging from 15 to 30 per 100,000 populations due to average low temperature. Brazil has a slightly higher incidence rate than European countries where 50 TB cases per 100,000 were reported, and China has 100 TB cases per 100,000 populations. The lowest TB incidence rate (5 per 100,000 populations) has been reported in the United States [31].

Currently, HIV is the major risk factor for developing TB [35] ranging from 20 to 30 [36]. Therefore, African countries having an increased rate of HIV infection also has a concurrent increase in the incidence of TB. Apart from HIV, other factors including malnutrition [37,38], smoking [39,40], indoor air pollution [40], and alcohol consumption [41,42] are common risk factors. Malnutrition and smoking are major contributors to TB. Several noncommunicable diseases like diabetes and obesity also have a significant impact on TB [43−45]. Excessive alcohol consumption among adults and diabetes is the leading risk factor in low- to middle-income countries [45]. Data suggested that 40% and 15% of the TB cases have been reported due to smoking [43] and diabetes [46] respectively, in India. In China, it has been estimated that 20%−50% incidence of TB will decline by

2033 by reducing the incidence of smoking and exposure to air pollution [47]. Correlation of silicosis with an incidence of TB has been well established [48]. Moreover, mental disorders and outside air pollution are thought to be risk factors for TB; however, this assumption has not been validated. Additionally, populations with weak immune defense are more prone to *M. tuberculosis* infection. Age is an important cognitive factor for the incidence of TB. Older age people and young ones are more prone to *M. tuberculosis* infection due to low immunity.

3.4 Approaches to disease control and elimination

Tuberculosis is a global health issue that needs to be controlled and requires extensive strategies to eliminate the disease from the root [49,50]. Elimination of tuberculosis signifies the achievement of incidences of tuberculosis infection in less than 1% per million populations [51]. Hence, the elimination of the global health issue requires advanced interventions along with standard control measures. One of the effective approaches to control and eliminate tuberculosis is through mass drug administration (MDA) [52,53]. It is a safe and inexpensive delivery of medicines, which is based on chemotherapy principles [54]. In this approach, without suspecting any individual diagnosis or infection status, a population or a subpopulation is provided with the treatment [55]. To eliminate such kinds of diseases, the World Health Organization has recommended mass drug administration as a useful approach [56]. The drugs administered under the mass drug administration technique have different mechanisms of action leading to different effects [57]. However, it has also been made evident that mass drug administration is effective only for a limited period and hence to achieve permanent elimination, it becomes necessary to administer the drugs repeatedly [54]. For effective disease control, the implementation of mass drug administration resides on several factors that include the number of rounds, treatment frequency, and timing [58]. The control, as well as the elimination of tuberculosis, also depends largely on epidemiological factors like transmission level and infection incorporation rates [59]. The national tuberculosis control programs should implement mass drug administration treatment twice in a year with maximum coverage of essential drugs. It also requires community participation by providing health education and necessary information regarding the disease to the people to improve coverage as well as to attain sustainability.

Considering these perspectives, the mass drug administration campaign for tuberculosis requires effective management of understated phases:

1. **Design phase**: The MDA campaign should focus on determining an MDA strategy for the target population.
2. **Planning and preparation phase**: In this phase, the microplan for obtaining demographic information needs to be gathered along with delivery strategies, the number of human resources required, logistical information, and pharmacovigilance plan.

3. **Implementation phase**: In this phase, the actual distribution of the treatment is focused.
4. **Monitoring and evaluation**: In this phase, the obstacles in the treatment process are identified to eliminate them with immediate action.

These above-mentioned steps are important for the effective execution of MDA programs. However, it is also noteworthy to mention that MDA programs should exclude special population groups including pregnant women and infants of less than 6 months old and under 5 kg body weights.

3.5 Next-generation vaccines, drugs, and diagnostics

To control the devastating disease, there is an urgent need to introduce novel vaccines and drugs [60,61]. The causal agent of tuberculosis, *Mycobacterium tuberculosis* (Mtb), remains a major health concern over the globe [62]. Bacilli Calmette-Guerin (BCG) has been used as a treatment for 90 years. However, it provides very limited protection from the disease. The evidence suggests that the BCG has the protective efficacy up to only 50%; however, it varies significantly due to factors such as geographic location and preexposure of mycobacterium [63].

In this context, the next generation of vaccines or subsequent development in existing drugs and vaccines is pivotal. Certain novel vaccines for tuberculosis include mycobacterial strains engineering derived vaccines, subunit vaccines, and recombinant BCG (r-BCG) that claims to combat the epidemic behavior of the disease [64].

The BCG strains are generally engineered to promote the immune responses that are done by gene insertions for Mtb antigens, host-resistant factors, and mammalian cytokines along with bacterial toxin-derived adjuvants insertion [65]. Apart from this, the manipulation of bacterial genes is also done to promote the antigen presentation as well as immune activation. For instance, VPM1002 has been regarded as a second-generation derivative that is an r-BCG and its engineering results in the expression of *L. monocytogenes* protein listeriolysin O (LLO) [66]. The evidence suggests that LLO expression in r-BCG results in increased autophagy along with apoptosis. It is also assumed that it helps in protecting a postexposure vaccine. The next-generation vaccines for VPM1002 are in the pipeline and are assumed as a promising drug candidate for the treatment of tuberculosis [67]. Some of its candidates are VPM1002 ΔnuoG, having increased protection and safety. Another candidate is BCG ΔureC:hly Δpdx1, or VPM1002 Δpdx1, generated with an attempt to further increase the protection as well as safety of vaccines and to make it suitable for immunocompromised individuals immunization such as infants and adults with HIV positive along with neonates that are HIV exposed and are at higher risk of developing tuberculosis [68]. On the other hand, the other vaccines like subunit vaccines were also supposed to have the potential of replacing BCG. However, the protection provided by subunit vaccines is less in comparison to living attenuated mycobacterial strains. If subunit vaccines are administered as boosters to a BCG prime, it might result in the increased efficacy of protection [69].

Additionally, the repertoire of the genomic information of MTB is beneficial for the identification of protein targets, which are essential for the survival of the bacteria [70]. Furthermore, looking at the recent advances in the field of structural proteomics in X-ray crystallography, NMR, Mass spectroscopy, and computational analysis, it is possible to elucidate the structural information of these protein drug targets [71−73]. Detailed structural knowledge has helped in better understanding of biosynthetic pathways and proteins involved in these pathways. For identification of a lead molecule, it will be advantageous to identify compounds, which bind and inhibit these protein drug targets. Inhibition of such proteins will further block an important biosynthetic pathway of the bacteria, which in turn will affect the survival of the bacteria [74].

3.6 New pathways as drug targets

Tuberculosis is still a foremost threat to global human health, and due to the manifestation of multiple drug-resistance (MDR) and widespread drug resistance (XDR), species of *Mtb* have led to a search for novel drug targets. The enlargement of the new diagnostic or novel discovery, distinctive therapeutic targets, new pathways, novel antimicrobial molecules, and recognition of innovative scaffolds are urgently needed to conflict the *Mtb* [75]. Despite this, a better knowledge of physiology, metabolic state, and mechanism of *Mtb* of the bacillus during its dormancy stage is an essential apprehension in hunt of drug targets in antagonism to nonreplicating persistent form (NRP) of the *Mtb*. The recognition of eventual drug targets adjacent to the dormant state of *Mtb* is significantly critical to accomplish the objective of the whole abolition for the restriction of anti-TB therapy [76]. The various metabolic pathways and their enzymes that could be potential drug targets are explained later:

3.6.1 Oxidative phosphorylation in *M. tuberculosis* as the target of antibacterial

Bedaquiline kills equally dormant and actively replicating mycobacteria by inhibiting mycobacterial adenosine triphosphate (ATP) synthase, a vital membrane-bound enzyme, interfering with energy production and distracting intracellular metabolism [77] and this opens a new door for targeting the pathogen employing oxidative phosphorylation system as a drug target. Oxidative phosphorylation system is enormously crucial for the survival of dormant *Mtb* in Latent *Mtb* infection because the intracellular concentration of ATP is 5−6 times lesser in hypoxic nonreplicating *Mtb* cells compared to aerobic replicating bacteria, making them precisely vulnerable to any extra depletion [78].

3.6.2 Drug targets important to sustain viability isocitrate lyase

The dormant state of *Mtb* is a nonreplicating persistence (NRP) form, characterized by slow growth and low metabolic state, and it is resistant to antitubercular (anti-TB) agents [76]. Due to the stress environment, *Mtb* cannot use the tricarboxylic acid (TCA) cycle for the generation of energy as the TCA cycle enzymes are downregulated. To compensate for the energy, *Mtb* uses an alternative pathway for the production of energy that is the glyoxylate cycle [79]. The conversion of isocitrate to succinate and glyoxylate, followed by the addition of acetyl-CoA to glyoxylate to form malate by malate synthase is catalyzed by the enzyme Isocitrate lyase (ICL). The ICL enzyme is vital for the survival of latent *Mtb*, and it is a potential drug target as the glyoxylate pathway is absent in the host [80].

3.6.3 Nitrate reduction pathways

To understand the pathogenesis of mycobacterium and its perseverance, the subsistence of indispensable nutrients in the host and their metabolism by the pathogen remains a fascinating area. In different mycobacterial species, nitrate metabolic pathways are operative under different physiological conditions that are helpful to recognize the biology of mycobacteria through the latent stage [81]. As the latent stage of *Mtb* is anaerobic and the reduction of nitrate to nitrite could generate the ATP that is indispensable to keep the pathogen alive, catalyzed by four genes namely narG, narH, narJ, and narI, clustered together as narGHJI in an operon, is a prospective drug target [82].

3.6.4 Mycobacterial lipase-LipY

Triacylglycerol (TAG), which is a triester of glycerol with a fatty acid, is the key energy storehouse of all eukaryotes including *Mtb*. The catabolism of the TAG generates twice the ATP than the same weight of carbohydrate or protein, and it is essential for the survival of dormant bacteria of *Mtb* during the latent stage [83]. The hydrolysis of the TAG is catalyzed by the enzyme Rv3097c (LipY) for the production of ATP in oxygen-depleted and nutrient starvation conditions [84]. Saxena et al. [85] identified the orlistat as the only known LipY inhibitor that showed the selective anti-TB activity toward dormant bacilli under hypoxic conditions and thus LipY is an appropriate drug target. Later on, Singh et al. [86] recognized the function of LipY in the virulence of *Mtb*.

3.6.5 Resuscitation-promoting factors

Resuscitation-promoting factor (Rpfs) proteins are peptidoglycan glycosidases accomplished of resuscitating dormant mycobacteria and have been found to play a role in the pathogenesis of tuberculosis. *Mtb* contains five Rpf genes: rpfA, rpfB, rpfC, rpfD, and rpfE encoding a protein Rv0867c, Rv1009, Rv1884c, Rv2389c, and Rv2450c that have a conserved RPF domain [87,88]. All the Rpfs play a crucial role in the resuscitation of dormant bacilli [89] and thus represent the attractive drug target.

3.6.6 NAD synthetase

Biosynthesis of nicotinamide adenine dinucleotide (NAD), an essential cofactor has an indispensable role, conserved in all mycobacterial genomes that can be potentially targeted to kill both replicating and nonreplicating *M. tuberculosis* strains. NaMN adenylyltransferase (NadD) and NAD synthetase (NadE) are the important enzymes of NAD biosynthesis also contributing toward dormancy conditions. Besides, NadD or NadE eradication blocks NAD synthesis, leading to depletion of the cofactor pool [90]. Previously, Boshoff et al. [91] identified that these two enzymes are desirable chemotherapeutic targets for active as well as latent TB.

3.6.7 Mycobactin biosynthesis

The *Mtb* bacterium needs an iron when it is inside an infected host; for this, it relies on the production of mycobactin (MBT). Due to the reliance on the iron, there is a great interest to interpret the mechanism of biosynthetic pathway MBT for antituberculosis drug development [92]. In MBT synthesis, the hydroxylation of lysine moiety is catalyzed by MBTG that is desirable for iron-chelating and thus it is an effective target in this pathway [93]. To accomplish the need of iron, obligatory by the host defensive system, bacteria have evolved iron attainment systems where small molecules known as siderophores are secreted, which combine extracellular iron and reabsorb through specific cell surface receptors [94,95]. The siderophore formed by the *M. tuberculosis* is the MBT class containing a salicylic acid-derived moiety. The MBT biogenesis is catalyzed by the 10-gene clusters that are designated as MBTA-J. Among the 10 genes, MBTB, MBTE, and MBTF are anticipated to be peptide synthetases, MBTI an isochorismate synthase that gives a salicylate activated by MBTA, and MBTG is a necessary hydroxylase [96], rendering these enzymes as appropriate drug targets.

3.6.8 MurD ligase (Rv2155c)—a potential broad-spectrum target

The biochemical pathway for peptidoglycan biosynthesis is one of the best sources of antibacterial targets [97]. Within this pathway, the MurD ligase is a suitable target for the growth of new classes of antibacterial agents. It belongs to the Mur CDEF ligase family having a molecular weight of 49.3 kDa, expressed in the cytoplasm as a monomer. MurD is overexpressed in its active form and homologous to MurD from *E. coli*. It catalyzes the formation of the peptide bond between UDP *N*-acetylmuramoyl-L-alanine and D-glutamate [98].

3.6.9 TMPK (EC 2.7.4.9) is the last specific enzyme in the pyrimidine biosynthetic pathway

Thymidylate kinase (EC 2.7.4.9) is the last specific enzyme in the pyrimidine biosynthetic pathway (TMPK, also called thymidine 5′-monophosphate kinase) and phosphorylates thymidine 5′-monophosphate (dTMP) to thymidine 5′-

diphosphate (dTDP) utilizing ATP as its preferred phosphoryl donor [99] that is essential for DNA replication. Thus, TMPKmt is critical for cell proliferation as well as for the survival of the *Mtb*, and thus a prospective target for developing new antituberculosis agents [100].

3.6.10 Rv0864

Rv0864 (MoaC2) from *M. tuberculosis* is one of the enzymes in the molybdenum cofactor (Moco) biosynthesis pathway. In concert with MoaA, MoaC is involved in the conversion of guanosine triphosphate to precursor Z, the first step in Moco synthesis [101]. In the primate model, a mutation in MoaC1 (Rv3111) of *M. tuberculosis* causes attenuation; therefore, the mechanism of the Moco pathway is essential to recognize which is still not apparent [102]. In a while, the crystal structure of the Moco-biosynthesis protein MoaC2 (Rv0864) was determined by Srivastava et al. [103]; it is found that it has a hexameric structure, homologous to *E. coli*. The crystal structure of Rv0864 gives new insights to design new inhibitors against MoaC2 protein.

3.6.11 Cytidylate kinase and polyphosphate kinase as drug targets

The Rv1712, Rv2984, Rv2194, Rv1311, Rv1305, and Rv1622c were formerly recognized as potential drug targets [104]. Initially, cytidylate kinase (Rv1712) and polyphosphate kinase (Rv2984) were identified through In silico and experimental studies [105,106] The Rv2984 is an impending drug target, experimentally verified by two of the studies [106,107]; whereas, for Rv1712, no PDB structures were identified. According to Khoshkholgh-Sima et al. [104], remaining proteins Rv2195, Rv1456c, and Rv2421c have also been identified as a potential drug target.

3.7 Concluding remarks

Tuberculosis affects people all over the world and contributes to substantial morbidity and death. A limited number of treatment options and extensive spread of MDR and XDR, the best therapy for tuberculosis will be an efficacious, safe, oral, and short-course combination treatment. To this end, novel chemicals targeting different parasites, biological processes, as well as novel drug delivery modes are being explored. A better understanding of pathophysiological and immunological complexities of the varied disease presentations and animal models that closely reflect human immune responses is a prerequisite to developing such a vaccine. The accessibility of genomes of several Mycobacterium species together with developments in the other omics fields has added to the pace of discovery of drug/vaccine targets. Although the strategic approaches and priority interventions may vary with the region, the total abolition of tuberculosis will depend on the mutual efforts of governments, scientific researchers, the pharmaceutical industry, and the vulnerable community.

References

[1] Cole S, Brosch R, Parkhill J, Garnier T, Churcher C, Harris D, Gordon SV, Eiglmeier K, Gas S, Barry III CE, Tekaia F. Deciphering the biology of *Mycobacterium tuberculosis* from the complete genome sequence. Nature June 1998; 393(6685):537.

[2] Eurosurveillance Editorial Team. WHO publishes Global tuberculosis report 2013. Euro Surveillance October 24, 2013;18(43):20615.

[3] Gandhi NR, Moll A, Sturm AW, Pawinski R, Govender T, Lalloo U, Zeller K, Andrews J, Friedland G. Extensively drug-resistant tuberculosis as a cause of death in patients co-infected with tuberculosis and HIV in a rural area of South Africa. The Lancet November 4, 2006;368(9547):1575−80.

[4] Zumla A, George A, Sharma V, Herbert RH, Oxley A, Oliver M. The WHO 2014 global tuberculosis report—further to go. The Lancet Global Health January 1, 2015;3(1):e10−2.

[5] . Falzon D, Jaramillo E, Schünemann HJ, Arentz M, Bauer M, Bayona J, Blanc L, Caminero JA, Daley CL, Duncombe C, Fitzpatrick C. WHO guidelines for the programmatic management of drug-resistant tuberculosis: 2011 update.

[6] Forget EJ, Menzies D. Adverse reactions to first-line antituberculosis drugs. Expert Opinion on Drug Safety March 1, 2006;5(2):231−49.

[7] Zumla AI, Gillespie SH, Hoelscher M, Philips PP, Cole ST, Abubakar I, McHugh TD, Schito M, Maeurer M, Nunn AJ. New antituberculosis drugs, regimens, and adjunct therapies: needs, advances, and future prospects. The Lancet Infectious Diseases April 1, 2014;14(4):327−40.

[8] Cox E, Laessig K. FDA approval of bedaquiline—the benefit−risk balance for drug-resistant tuberculosis. New England Journal of Medicine August 21, 2014;371(8): 689−91.

[9] Ryan NJ, Lo JH. Delamanid: first global approval. Drugs June 1, 2014;74(9):1041−5.

[10] Kent WJ, Sugnet CW, Furey TS, Roskin KM, Pringle TH, Zahler AM, Haussler D. The human genome browser at UCSC. Genome Research June 1, 2002;12(6):996−1006.

[11] Galperin MY, Koonin EV. Searching for drug targets in microbial genomes. Current Opinion in Biotechnology December 1, 1999;10(6):571−8.

[12] Pande TR, Cohen C, Pai M, Ahmad Khan F. Computer-aided detection of pulmonary tuberculosis on digital chest radiographs: a systematic review. International Journal of Tuberculosis and Lung Disease September 1, 2016;20(9):1226−30.

[13] World Health Organization. Chest radiography in tuberculosis detection: summary of current WHO recommendations and guidance on programmatic approaches. World Health Organization; 2016.

[14] Lawn SD, Gupta-Wright A. Detection of galactomannan (LAM) in urine is indicative of disseminated TB with renal involvement in patients living with HIV and advanced immunodeficiency: evidence and implications. Transactions of the Royal Society of Tropical Medicine and Hygiene February 16, 2016;110(3):180−5.

[15] Lawn SD. Point-of-care detection of galactomannan (LAM) in urine for diagnosis of HIV-associated tuberculosis: a state of the art review. BMC Infectious Diseases December 2012;12(1):103.

[16] Gupta-Wright A, Corbett EL, van Oosterhout JJ, Wilson D, Grint D, Alufandika-Moyo M, Peters JA, Chiume L, Flach C, Lawn SD, Fielding K. Rapid urine-based screening for tuberculosis in HIV-positive patients admitted to hospital in Africa (STAMP): a pragmatic, multicentre, parallel-group, double-blind, randomised controlled trial. The Lancet July 28, 2018;392(10144):292−301.

[17] Sigal GB, Pinter Λ, Lowary TL, Kawasaki M, Li A, Mathew A, Tsionsky M, Zheng RB, Plisova T, Shen K, Katsuragi K. A novel sensitive immunoassay targeting the MTX-Lipoarabinomannan epitope meets the WHO's performance target for Tuberculosis diagnosis. Journal of Clinical Microbiology September 26, 2018;56. JCM-01338.

[18] Paris L, Magni R, Zaidi F, Araujo R, Saini N, Harpole M, Coronel J, Kirwan DE, Steinberg H, Gilman RH, Petricoin EF. Urine lipoarabinomannan glycan in HIV-negative patients with pulmonary tuberculosis correlates with disease severity. Science Translational Medicine December 13, 2017;9(420):eaal2807.

[19] Kana BD, Gordhan BG, Downing KJ, Sung N, Vostroktunova G, Machowski EE, Tsenova L, Young M, Kaprelyants A, Kaplan G, Mizrahi V. The resuscitation-promoting factors of *Mycobacterium tuberculosis* are required for virulence and resuscitation from dormancy but are collectively dispensable for growth in vitro. Molecular Microbiology February 2008;67(3):672−84.

[20] Cazabon D, Pande T, Kik S, Van Gemert W, Sohn H, Denkinger C, Qin ZZ, Waning B, Pai M. Market penetration of Xpert *MTB*/RIF in high tuberculosis burden countries: a trend analysis from 2014−2016. Gates Open Research July 25, 2018;2.

[21] Albert H, Nathavitharana RR, Isaacs C, Pai M, Denkinger CM, Boehme CC. Development, roll-out and impact of Xpert *MTB*/RIF for tuberculosis: what lessons have we learnt and how can we do better? European Respiratory Journal August 1, 2016; 48(2):516−25.

[22] Xie YL, Chakravorty S, Armstrong DT, Hall SL, Via LE, Song T, Yuan X, Mo X, Zhu H, Xu P, Gao Q. Evaluation of a rapid molecular drug-susceptibility test for tuberculosis. New England Journal of Medicine September 14, 2017;377(11): 1043−54.

[23] Yadav RN, Singh BK, Sharma SK, Sharma R, Soneja M, Sreenivas V, Myneedu VP, Hanif M, Kumar A, Sachdeva KS, Paramasivan CN. Correction: comparative evaluation of GenoType *MTB*DRplus line probe assay with solid culture method in early diagnosis of multidrug resistant tuberculosis (MDR-TB) at a tertiary care centre in India. PLoS One 2013;8(10).

[24] Singh BK, Sharma SK, Sharma R, Sreenivas V, Myneedu VP, Kohli M, Bhasin D, Sarin S. Diagnostic utility of a line probe assay for multidrug resistant-TB in smear-negative pulmonary tuberculosis. PLoS One August 22, 2017;12(8):e0182988.

[25] . Hain lifescience GmbH. Nehren, Germany. GenoType MTBDRplus, version 2.0 [product insert]. http://www.hain-lifescience.de/en/instructions-for-use.html.

[26] CRyPTIC Consortium and the 100,000 Genomes Project. Prediction of susceptibility to first-line tuberculosis drugs by DNA sequencing. New England Journal of Medicine October 11, 2018;379(15):1403−15.

[27] Walker TM, Ip CL, Harrell RH, Evans JT, Kapatai G, Dedicoat MJ, Eyre DW, Wilson DJ, Hawkey PM, Crook DW, Parkhill J. Whole-genome sequencing to delineate *Mycobacterium tuberculosis* outbreaks: a retrospective observational study. The Lancet Infectious Diseases February 1, 2013;13(2):137−46.

[28] Sulis G, Roggi A, Matteelli A, Raviglione MC. Tuberculosis: epidemiology and control. Mediterranean Journal of Hematology and Infectious Diseases 2014;6(1).

[29] Singh BK, Sharma SK, Sharma R, Sreenivas V, Myneedu VP, Kohli M, Sarin S. Diagnostic utility of a line probe assay for multidrug resistant-TB in smear-negative pulmonary tuberculosis. PLoS One 2017;12(8).

[30] Gupta S, Kakkar V. Recent technological advancements in tuberculosis diagnostics—a review. Biosensors and Bioelectronics 2018;115:14—29.

[31] World Health Organisation. Global tuberculosis report 2018. France: World Health Organization; 2018.

[32] Holloway KL, Henneberg RJ, de Barros Lopes M, Henneberg M. Evolution of human tuberculosis: a systematic review and meta-analysis of paleopathological evidence. HOMO-Journal of Comparative Human Biology December 1, 2011;62(6):402—58.

[33] Lienhardt C, Glaziou P, Uplekar M, Lönnroth K, Getahun H, Raviglione M. Global tuberculosis control: lessons learnt and future prospects. Nature Reviews Microbiology June 2012;10(6):407.

[34] Vynnycky E, Fine PE. Lifetime risks, incubation period, and serial interval of tuberculosis. American Journal of Epidemiology August 1, 2000;152(3):247—63.

[35] Havlir DV, Barnes PF. Tuberculosis in patients with human immunodeficiency virus infection. New England Journal of Medicine February 4, 1999;340(5):367—73.

[36] World Health Organization. Global tuberculosis control: a short update to the 2009 report. Geneva: World Health Organization; 2009.

[37] Cegielski JP, McMurray DN. The relationship between malnutrition and tuberculosis: evidence from studies in humans and experimental animals. International Journal of Tuberculosis and Lung Disease March 1, 2004;8(3):286—98.

[38] . Lönnroth K, Williams BG, Cegielski P, Dye C. A homogeneous dose-response relationship between body-mass index and tuberculosis incidence.

[39] Slama K, Chiang CY, Enarson DA, Hassmiller K, Fanning A, Gupta P, Ray C. Tobacco and tuberculosis: a qualitative systematic review and meta-analysis. International Journal of Tuberculosis and Lung Disease October 1, 2007;11(10):1049—61.

[40] Lin HH, Ezzati M, Murray M. Tobacco smoke, indoor air pollution and tuberculosis: a systematic review and meta-analysis. PLoS Medicine January 16, 2007;4(1):e20.

[41] Lönnroth K, Williams BG, Stadlin S, Jaramillo E, Dye C. Alcohol use as a risk factor for tuberculosis—a systematic review. BMC Public Health December 2008;8(1):289.

[42] Rehm J, Samokhvalov AV, Neuman MG, Room R, Parry C, Lönnroth K, Patra J, Poznyak V, Popova S. The association between alcohol use, alcohol use disorders and tuberculosis (TB). A systematic review. BMC Public Health December 2009; 9(1):450.

[43] Stevenson CR, Critchley JA, Forouhi NG, Roglic G, Williams BG, Dye C, Unwin NC. Diabetes and the risk of tuberculosis: a neglected threat to public health? Chronic Illness September 2007;3(3):228—45.

[44] Jeon CY, Murray MB. Correction: diabetes mellitus increases the risk of active tuberculosis: a systematic review of 13 observational studies. PLoS Medicine August 26, 2008;5(8):e181.

[45] Dooley KE, Chaisson RE. Tuberculosis and diabetes mellitus: convergence of two epidemics. The Lancet Infectious Diseases December 1, 2009;9(12):737—46.

[46] Stevenson CR, Forouhi NG, Roglic G, Williams BG, Lauer JA, Dye C, Unwin N. Diabetes and tuberculosis: the impact of the diabetes epidemic on tuberculosis incidence. BMC Public Health December 2007;7(1):234.

[47] Lin HH, Murray M, Cohen T, Colijn C, Ezzati M. Effects of smoking and solid-fuel use on COPD, lung cancer, and tuberculosis in China: a time-based, multiple risk factor, modelling study. The Lancet October 25, 2008;372(9648):1473−83.

[48] Barboza CE, Winter DH, Seiscento M, Santos UD, Terra Filho M. Tuberculosis and silicosis: epidemiology, diagnosis and chemoprophylaxis. Jornal Brasileiro de Pneumologia November 2008;34(11):959−66.

[49] Lönnroth K, Migliori GB, Abubakar I, D'Ambrosio L, De Vries G, Diel R, Douglas P, Falzon D, Gaudreau MA, Goletti D, Ochoa ER. Towards tuberculosis elimination: an action framework for low-incidence countries. European Respiratory Journal April 1, 2015;45(4):928−52.

[50] Raviglione MC, Smith IM. XDR tuberculosis—implications for global public health. New England Journal of Medicine February 15, 2007;356(7):656−9.

[51] Dye C, Williams BG. Eliminating human tuberculosis in the twenty-first century. Journal of the Royal Society Interface August 9, 2007;5(23):653−62.

[52] Hotez P. Mass drug administration and integrated control for the world's high-prevalence neglected tropical diseases. Clinical Pharmacology and Therapeutics June 2009;85(6):659−64.

[53] Raviglione MC, Pio A. Evolution of WHO policies for tuberculosis control, 1948−2001. The Lancet March 2, 2002;359(9308):775−80.

[54] Webster JP, Molyneux DH, Hotez PJ, Fenwick A. The contribution of mass drug administration to global health: past, present and future. Philosophical Transactions of the Royal Society B: Biological Sciences June 19, 2014;369(1645):20130434.

[55] Newby G, Hwang J, Koita K, Chen I, Greenwood B, Von Seidlein L, Shanks GD, Slutsker L, Kachur SP, Wegbreit J, Ippolito MM. Review of mass drug administration for malaria and its operational challenges. The American Journal of Tropical Medicine and Hygiene July 8, 2015;93(1):125−34.

[56] World Health Organization. Consolidated guidelines on the use of antiretroviral drugs for treating and preventing HIV infection: recommendations for a public health approach. World Health Organization; 2016.

[57] Bandura A. Social cognitive theory of mass communication. Media Effects January 13, 2009:110−40. Routledge.

[58] Beaglehole R, Bonita R, Horton R, Adams C, Alleyne G, Asaria P, Baugh V, Bekedam H, Billo N, Casswell S, Cecchini M. Priority actions for the non-communicable disease crisis. The Lancet April 23, 2011;377(9775):1438−47.

[59] Lönnroth K, Raviglione M. Global epidemiology of tuberculosis: prospects for control. Seminars in Respiratory and Critical Care Medicine October 2008;29(5):481−91. © Thieme Medical Publishers.

[60] Kaufmann SH, Hussey G, Lambert PH. New vaccines for tuberculosis. The Lancet June 12, 2010;375(9731):2110−9.

[61] Koul A, Arnoult E, Lounis N, Guillemont J, Andries K. The challenge of new drug discovery for tuberculosis. Nature January 2011;469(7331):483.

[62] Dara M, Acosta CD, Melchers NV, Al-Darraji HA, Chorgoliani D, Reyes H, Centis R, Sotgiu G, D'Ambrosio L, Chadha SS, Migliori GB. Tuberculosis control in prisons: current situation and research gaps. International Journal of Infectious Diseases March 1, 2015;32:111−7.

[63] Nieuwenhuizen NE, Kaufmann SH. Next-generation vaccines based on Bacille Calmette−Guérin. Frontiers in Immunology February 5, 2018;9:121.

[64] Andersen P, Woodworth JS. Tuberculosis vaccines—rethinking the current paradigm. Trends in Immunology August 1, 2014;35(8):387—95.

[65] Nor NM, Musa M. Approaches towards the development of a vaccine against tuberculosis: recombinant BCG and DNA vaccine. Tuberculosis January 1, 2004;84(1—2): 102—9.

[66] Kaufmann SH, Cotton MF, Eisele B, Gengenbacher M, Grode L, Hesseling AC, Walzl G. The BCG replacement vaccine VPM1002: from drawing board to clinical trial. Expert Review of Vaccines May 1, 2014;13(5):619—30.

[67] Kaufmann SH, Lange C, Rao M, Balaji KN, Lotze M, Schito M, Zumla AI, Maeurer M. Progress in tuberculosis vaccine development and host-directed therapies—a state of the art review. The Lancet Respiratory Medicine April 1, 2014; 2(4):301—20.

[68] Hawkridge T, Mahomed H. Prospects for a new, safer and more effective TB vaccine. Paediatric Respiratory Reviews March 1, 2011;12(1):46—51.

[69] Kaufmann SH. Fact and fiction in tuberculosis vaccine research: 10 years later. The Lancet Infectious Diseases August 1, 2011;11(8):633—40.

[70] Smith DR. Microbial pathogen genomes—new strategies for identifying therapeutics and vaccine targets. Trends in Biotechnology August 1, 1996;14(8):290—3.

[71] Ilari A, Savino C. Protein structure determination by x-ray crystallography. Bioinformatics 2008:63—87. Humana Press.

[72] Cavalli A, Salvatella X, Dobson CM, Vendruscolo M. Protein structure determination from NMR chemical shifts. Proceedings of the National Academy of Sciences United States of America June 5, 2007;104(23):9615—20.

[73] Singh P, Panchaud A, Goodlett DR. Chemical cross-linking and mass spectrometry as a low-resolution protein structure determination technique. Analytical Chemistry April 1, 2010;82(7):2636—42.

[74] Zhang Y, Post-Martens K, Denkin S. New drug candidates and therapeutic targets for tuberculosis therapy. Drug Discovery Today January 1, 2006;11(1—2):21—7.

[75] Bald D, Villellas C, Lu P, Koul A. Targeting energy metabolism in *Mycobacterium tuberculosis*, a new paradigm in antimycobacterial drug discovery. mBio May 3, 2017;8(2). e00272-17.

[76] Gupta VK, Kumar MM, Singh D, Bisht D, Sharma S. Drug targets in dormant *Mycobacterium tuberculosis*: can the conquest against tuberculosis become a reality? Infectious Diseases February 1, 2018;50(2):81—94.

[77] Field SK. Bedaquiline for the treatment of multidrug-resistant tuberculosis: great promise or disappointment? Therapeutic Advances in Chronic Disease July 2015; 6(4):170—84.

[78] Machado D, Pires D, Perdigão J, Couto I, Portugal I, Martins M, Amaral L, Anes E, Viveiros M. Ion channel blockers as antimicrobial agents, efflux inhibitors, and enhancers of macrophage killing activity against drug resistant *Mycobacterium tuberculosis*. PLoS One February 26, 2016;11(2):e0149326.

[79] Sharma V, Sharma S, zu Bentrup KH, McKinney JD, Russell DG, Jacobs Jr WR, Sacchettini JC. Structure of isocitrate lyase, a persistence factor of *Mycobacterium tuberculosis*. Nature Structural and Molecular Biology August 2000;7(8):663.

[80] McKinney JD, Zu Bentrup KH, Muñoz-Elías EJ, Miczak A, Chen B, Chan WT, Swenson D, Sacchettini JC, Jacobs Jr WR, Russell DG. Persistence of *Mycobacterium tuberculosis* in macrophages and mice requires the glyoxylate shunt enzyme isocitrate lyase. Nature August 2000;406(6797):735.

[81] zu Bentrup KH, Russell DG. Mycobacterial persistence: adaptation to a changing environment. Trends in Microbiology December 1, 2001;9(12):597−605.

[82] Sohaskey CD. Nitrate enhances the survival of *Mycobacterium tuberculosis* during inhibition of respiration. Journal of Bacteriology April 15, 2008;190(8):2981−6.

[83] Deb C, Daniel J, Sirakova TD, Abomoelak B, Dubey VS, Kolattukudy PE. A novel lipase belonging to the hormone-sensitive lipase family induced under starvation to utilize stored triacylglycerol in *Mycobacterium tuberculosis*. Journal of Biological Chemistry February 17, 2006;281(7):3866−75.

[84] Deb C, Lee CM, Dubey VS, Daniel J, Abomoelak B, Sirakova TD, Pawar S, Rogers L, Kolattukudy PE. A novel in vitro multiple-stress dormancy model for *Mycobacterium tuberculosis* generates a lipid-loaded, drug-tolerant, dormant pathogen. PLoS One June 29, 2009;4(6):e6077.

[85] Saxena AK, Roy KK, Singh S, Vishnoi SP, Kumar A, Kashyap VK, Kremer L, Srivastava R, Srivastava BS. Identification and characterisation of small-molecule inhibitors of Rv3097c-encoded lipase (LipY) of *Mycobacterium tuberculosis* that selectively inhibit growth of bacilli in hypoxia. International Journal of Antimicrobial Agents July 1, 2013;42(1):27−35.

[86] Singh VK, Srivastava M, Dasgupta A, Singh MP, Srivastava R, Srivastava BS. Increased virulence of *Mycobacterium tuberculosis* H37Rv overexpressing LipY in a murine model. Tuberculosis May 1, 2014;94(3):252−61.

[87] Downing KJ, Mischenko VV, Shleeva MO, Young DI, Young M, Kaprelyants AS, Apt AS, Mizrahi V. Mutants of *Mycobacterium tuberculosis* lacking three of the five rpf-like genes are defective for growth in vivo and for resuscitation in vitro. Infection and Immunity May 1, 2005;73(5):3038−43.

[88] Hett EC, Chao MC, Steyn AJ, Fortune SM, Deng LL, Rubin EJ. A partner for the resuscitation-promoting factors of *Mycobacterium tuberculosis*. Molecular Microbiology November 2007;66(3):658−68.

[89] Uhía I, Krishnan N, Robertson BD. Characterising resuscitation promoting factor fluorescent-fusions in mycobacteria. BMC Microbiology December 2018;18(1):30.

[90] Rodionova IA, Schuster BM, Guinn KM, Sorci L, Scott DA, Li X, Kheterpal I, Shoen C, Cynamon M, Locher C, Rubin EJ. Metabolic and bactericidal effects of targeted suppression of NadD and NadE enzymes in mycobacteria. mBio February 28, 2014;5(1). e00747-13.

[91] Boshoff HI, Xu X, Tahlan K, Dowd CS, Pethe K, Camacho LR, Park TH, Yun CS, Schnappinger D, Ehrt S, Williams KJ. Biosynthesis and recycling of nicotinamide cofactors in *Mycobacterium tuberculosis* an essential role for nad in nonreplicating bacilli. Journal of Biological Chemistry July 11, 2008;283(28):19329−41.

[92] McMahon MD, Rush JS, Thomas MG. Analyses of MbtB, MbtE, and MbtF suggest revisions to the mycobactin biosynthesis pathway in *Mycobacterium tuberculosis*. Journal of Bacteriology June 1, 2012;194(11):2809−18.

[93] Moody DB, Young DC, Cheng TY, Rosat JP, Roura-mir C, O'Connor PB, Zajonc DM, Walz A, Miller MJ, Levery SB, Wilson IA. T cell activation by lipopeptide antigens. Science January 23, 2004;303(5657):527−31.

[94] Braun V, Hantke K, Koester W. Bacterial iron transport: mechanisms, genetics, and regulation. Metal Ions in Biological Systems January 1, 1998;35:67−146.

[95] Byers BR, Arceneaux JE. Microbial iron transport: iron acquisition by pathogenic microorganisms. Metal Ions in Biological Systems 1998;35:37.

[96] Quadri LE, Sello J, Keating TA, Weinreb PH, Walsh CT. Identification of a *Mycobacterium tuberculosis* gene cluster encoding the biosynthetic enzymes for assembly of the virulence-conferring siderophore mycobactin. Chemistry and Biology November 1, 1998;5(11):631−45.

[97] Kouidmi I, Levesque RC, Paradis-Bleau C. The biology of Mur ligases as an antibacterial target. Molecular Microbiology October 2014;94(2):242−53.

[98] Crick DC, Mahapatra S, Brennan PJ. Biosynthesis of the arabinogalactan-peptidoglycan complex of *Mycobacterium tuberculosis*. Glycobiology September 1, 2001;11(9). 107R-18R.

[99] Sassetti CM, Boyd DH, Rubin EJ. Genes required for mycobacterial growth defined by high density mutagenesis. Molecular Microbiology April 2003;48(1):77−84.

[100] Gasse C, Huteau V, Douguet D, Munier-Lehmann H, Pochet S. A new family of inhibitors of *Mycobacterium tuberculosis* thymidine monophosphate kinase. Nucleosides, Nucleotides and Nucleic Acids November 26, 2007;26(8−9):1057−61.

[101] Wuebbens MM, Rajagopalan KV. Investigation of the early steps of molybdopterin biosynthesis in *Escherichia coli* through the use of in vivo labeling studies. Journal of Biological Chemistry January 20, 1995;270(3):1082−7.

[102] Dutta NK, Mehra S, Didier PJ, Roy CJ, Doyle LA, Alvarez X, Ratterree M, Be NA, Lamichhane G, Jain SK, Lacey MR. Genetic requirements for the survival of tubercle bacilli in primates. The Journal of Infectious Diseases June 1, 2010;201(11):1743−52.

[103] Srivastava VK, Srivastava S, Arora A, Pratap JV. Structural insights into putative molybdenum cofactor biosynthesis protein C (MoaC2) from *Mycobacterium tuberculosis* H37Rv. PLoS One March 19, 2013;8(3):e58333.

[104] Khoshkholgh-Sima B, Sardari S, Mobarakeh JI, Khavari-Nejad RA. An in silico approach for prioritizing drug targets in metabolic pathway of *Mycobacterium tuberculosis*. World Academy of Science, Engineering, and Technology. International Journal of Pharmacological and Pharmaceutical Sciences November 25, 2011;5: 613−6.

[105] Caceres RA, Timmers LF, Vivan AL, Schneider CZ, Basso LA, De Azevedo WF, Santos DS. Molecular modeling and dynamics studies of cytidylate kinase from *Mycobacterium tuberculosis* H37Rv. Journal of Molecular Modeling May 1, 2008;14(5): 427−34.

[106] Jagannathan V, Kaur P, Datta S. Polyphosphate kinase from *M. tuberculosis*: an interconnect between the genetic and biochemical role. PLoS One December 15, 2010; 5(12):e14336.

[107] Russell DG. Phagosomes, fatty acids and tuberculosis. Nature Cell Biology September 2003;5(9):776.

Lymphatic filariasis

Sivapong Sungpradit[1,2], Vivornpun Sanprasert[3,4]

[1]*Assistant Professor, Department of Pre-clinic and Applied Animal Science, Faculty of Veterinary Science, Mahidol University, Phutthamonthon, Nakhon Pathom, Thailand;* [2]*The Monitoring and Surveillance Center for Zoonotic Diseases in Wildlife and Exotic Animals, Faculty of Veterinary Science, Mahidol University, Phutthamonthon, Nakhon Pathom, Thailand;* [3]*Assistant Professor, Department of Parasitology, Faculty of Medicine, Chulalongkorn University, Pathumwan, Bangkok, Thailand;* [4]*Lymphatic Filariasis and Tropical Medicine Research Unit, Chulalongkorn Medical Research Center, Faculty of Medicine, Chulalongkorn University, Pathumwan, Bangkok, Thailand*

4.1 Introduction

Lymphatic filariasis, known as elephantiasis, is caused by mosquito-transmitted lymphatic filarial parasites, including *Wuchereria bancrofti, Brugia malayi,* and *B. timori.* The majority of the disease is caused by *W. bancrofti* accounting for 90% of the cases, and the minority accounting for 10% by *B. malayi,* and 0.67% by *B. timori* [1,2]. Although *W. bancrofti* is widely distributed in many tropical and subtropical areas, *Brugia* spp. is more restricted to South East Asian countries. *B. timori* is found in Timor and its surrounding islands [3,4]. Although lymphatic filariasis is not a fatal disease, it is responsible for considerable suffering, disability, and abnormality. In 1995, lymphatic filariasis is ranked by the World Health Organization (WHO) as the world's second leading cause of permanent and long-term disability [2,5]. In terms of "Disability Adjusted Life Years" (DALYs: the number of healthy years of life lost due to premature death and disability), lymphatic filariasis is responsible for 5.8 million DALYs lost annually [6], ranking third among the special program for research and training in Tropical Diseases Research, after malaria and tuberculosis [6].

The International Task Force on Disease Eradication (ITFDE) has identified lymphatic filariasis as one of the six potentially eradicable infectious diseases [5]. Following the World Health Assembly resolution of 1997 urging the WHO to eliminate lymphatic filariasis as a public health problem, WHO and interested partners from both the public and private sectors outlined the plan to achieve this goal. Lymphatic filariasis has been targeted for global elimination as a public health problem by the year 2020 [7]. The two principal elimination strategies employed by the Global Program for Elimination of Lymphatic Filariasis (GPELF) are 1) to interrupt transmission of infection and 2) to alleviate and prevent the suffering and disability caused by the disease.

Molecular Advancements in Tropical Diseases Drug Discovery. https://doi.org/10.1016/B978-0-12-821202-8.00004-9

4.1.1 Life cycle

The life cycle of filarial nematodes is biphasic where larval development takes place in mosquito vectors (intermediate host), while adult development takes place in humans (definitive host). The infection is transmitted by the bite of infected mosquitoes. During a blood meal, the infective larvae (third-stage larvae; L3s) penetrate the bite wound, passing to the lymphatic vessels and lymph nodes where they develop into adult worms. After mating, the female worm can produce about 50,000 microfilariae per day. Then, millions of microfilariae are periodically released into the host's blood circulation from the lymphatics at a certain time depending on the species of the parasites. Adult lymphatic filarial parasites have a life span of 5—10 years, while the microfilariae can live longer for 6—12 months [8]. When mosquitoes bite the infected individuals, microfilariae are ingested and shed their sheath, penetrate the stomach wall, and migrate to the thoracic muscles. Then, they change into the first-stage larvae and subsequently develop into the mature infective larvae. The infective larvae migrate to the mosquito's proboscis, from where they will be transferred to other people via next mosquito bites (Fig. 4.1).

Infection is usually acquired during childhood without any clinical symptoms but can cause an impaired lymphatic system [9]. Adult worms in lymphatic tissues induce pathological changes, including dysfunction or inflammatory damage of lymphatics, dilatation of lymphatics leading to the thickening and obstruction of the lymphatic vessel walls, causing the chronic irreversible pathology or

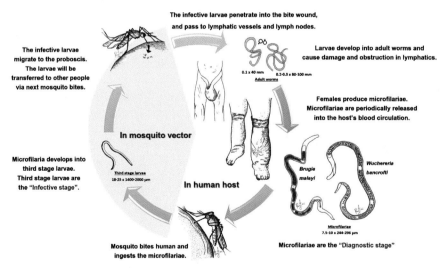

FIGURE 4.1

Lymphatic filariasis life cycle.

elephantiasis. The pathogenesis of lymphatic filariasis is thought to be caused by adult worms, host immune responses, and secondary bacterial infections [10]. The disease manifestations of lymphatic filariasis vary from subclinical (or asymptomatic) to clinical symptoms. Generally, most infected individuals develop an infection with circulating microfilariae. Other patients who carry the adult worms may be amicrofilaremic and asymptomatic, or have acute manifestations (lymphangitis, adenolymphangitis) and chronic manifestations (hematuria, hydrocele, chylocele, chyluria, lymphedema, and elephantiasis) [11]. Filarial infection can rarely manifest as tropical pulmonary eosinophilia [12]. Atypical manifestations can also occur, such as filarial arthritis, nephritic syndrome, and glomerulonephritis [13]. Men are more susceptible to infection and to chronic pathology than women [14]. However, gender is not an important influence on the incidence of acute manifestations in lymphatic filariasis. In endemic areas, some population remains uninfected despite exposure to the parasites to the same degree as the rest of the population. These individuals have been termed as "endemic normal." This usually depends on the sensitivity of tools to define infection status. Moreover, the definition of endemic normals has undergone many changes. The true prevalence of endemic normals needs to be reexamined with the advent of newer techniques for the diagnosis of subclinical infections.

In the 1970s, it has been discovered that filarial nematodes themselves harbor the intracellular bacteria *Wolbachia* [15]. These bacteriae are found widespread in arthropods as well as in filarial nematodes [16]. The association between *Wolbachia* and filarial nematodes is obligatory mutualism [17]. *Wolbachia* requires filarial nematode for normal larval development and reproduction [18]. The discovery of *Wolbachia* has encouraged researchers to develop them as potential targets for new chemotherapy, diagnosis, as well as identify immunomodulatory antigens for understanding the pathogenesis of the disease. Genomics and proteomics analysis may be helpful for the identification of the candidate molecules.

4.2 Detection of infection

The diagnosis of lymphatic filariasis is usually difficult as it involves the microscopic detection of the microfilariae in blood collected during the night time. It is the conventional method and remains the definite diagnosis of lymphatic filariasis. The simplest method for the detection of microfilariae is fresh blood smear using finger prick to observe the motility of microfilariae under a microscope. However, this method cannot identify the species of parasite. Giemsa's stain of thick and thin blood films is the most common method for the detection of microfilariae in the blood samples. However, both fresh blood smear and Giemsa's staining methods have the disadvantage of low sensitivity. More than 20 Mf/mL of blood is required for a reliable result by blood smear tests. The levels of microfilariae in finger-prick blood samples have been reported to be higher than the levels in venous blood samples [19]. The sensitivity of microfilariae detection can be improved by concentration techniques using a larger volume of blood samples, such as Knott's test, membrane filtration test. Moreover, the expert parasitologist is needed to distinguish microfilariae of *W. bancrofti* and *B. malayi* from other filarial parasites, especially in the coendemic areas of *Loa loa*, *Mansonella* spp., and *Onchocerca volvulus*.

Another limitation of microfilariae detection is the periodicity of the microfilariae strains, limiting the time of their appearance in blood circulation, leading to limiting the time for blood collection. The ability to detect circulating microfilariae is therefore highly dependent on the time of blood collection and volume of blood examined. The levels of microfilariae in blood circulation do not certainly correlate with adult worm burden in lymphatics. Microfilaremia is also a dynamic process, microfilaremic patients becoming amicrofilaremics still retain adult worms in lymphatic vessels [20]. It is, therefore, necessary to monitor the active infection and confirm that all infected individuals receive proper treatment with diethylcarbamazine citrate (DEC). Therefore, a method for quantifying the adult worm burden independent of microfilaraemia status is required to be developed, especially in individuals in endemic areas who are continuously exposed to the parasites and in patients with chronic disease who are usually amicrofilaremics.

The diagnostic approach has reformed by the development of circulating filarial antigen (CFA) detection assays. The development of a CFA detection method has been very useful not only for the diagnosis of the disease but also for monitoring the efficacy of the treatments as well as of the control program launched by WHO. These CFA assays demonstrate high sensitivity, specificity, and simplicity of use. Blood samples obtained from a single finger prick can be used and can be taken at any time of the day even in individuals with the nocturnally periodic infection. Unfortunately, two CFA assays commercially available to date are only specific to bancroftian filariasis. These assays have been developed based on two monoclonal antibodies, including Og4C3 mAb and AD12 mAb [21,22].

The Og4C3 mAb is a murine monoclonal antibody against nonphosphocholine antigen of *Onchocerca gibsoni* [22]. The Og4C3 ELISA or Trop Bio Og4C3 Antigen test was the first commercially available for the diagnosis of bancroftian filariasis. It is a quantitative test with high specificity (99%−100%) [22]. Although the sensitivity of Og4C3 ELISA is almost 100% in microfilaremic patients, it declines to 72%−75% in amicrofilaremics and patients with low-level microfilaremia [23]. The prevalence assessed by Og4C3 ELISA is about 5−20-fold higher than by microfilariae detection [24−27]. This is particularly important for amicrofilaremic patients who have been missed by microscopic methods.

The Binax NOW Filariasis ICT Card Test is an immunochromatographic test (ICT) for the qualitative detection of CFA from *W. bancrofti* using AD12 mAb. This antibody recognizes a phosphocholine antigen (200 kDa) secreted from *W. bancrofti* adult worms [21,28]. From the animal model, AD12 mAb can first detect *W. bancrofti* within 1 month of the onset of microfilaremia [29]. The ICT card test is a rapid test with high specificity (95%−100%). Cross-reactivity of the ICT card test with *Loa loa* and *O. ochengi* has been reported [30,31]. Moreover, the sensitivity of ICT has been reported to be lower than Og4C3 ELISA [25,27,32]. Capillary whole blood samples collected from the finger stick can be used and no need of any instrument or reagents. Therefore, the use of the ICT card test has prompted a suitable tool for the mapping of filariasis endemicity and has been used by the GPELF for more than 10 years. Indeed, this diagnostic tool showed that the amount of filariasis had been underestimated by microscopy alone. For the eradication campaign to be successful, public health departments must have accurate endemicity data. The Alere filariasis test strip (FTS) is the new filarial antigen test using the same mAb. This strip detected 26.5% more antigenemia patients and had better stability in a field study than the reference Binax Now card test [33]. The Alere FTS is therefore approved for use as a diagnostic tool for bancroftian filariasis in the GPELF.

Not only being a suitable tool for the mapping of filariasis endemicity, CFA assays are also useful for posttreatment assessment. A decrease of CFA levels indicates the macrofilaricidal efficacy of chemotherapy. However, CFA levels are still detected at 4 months after treatment [34]. Therefore, the treatment follow-up using CFA assays should be performed at least 4 months after treatment. Because CFA detection assays are now considered the gold standard for the diagnosis of bancroftian filariasis, the development of an antigen detection test for the diagnosis of brugian filariasis will be undoubtedly very useful to the public health programs of the Southeast Asian communities afflicted by this parasite. However, the issue of cost remains a concern in developing countries, especially in large-scale screening. The cost-effective, simple, and reliable serological methods are required to allow diagnosis and to gain information about the prevalence of the disease. Moreover, until now, there is no antigen detection assay available for the diagnosis of brugian filariasis. Several serological assays have been therefore developed to diagnose both bancroftian and brugian filariasis.

Serological assays based on the detection of antifilarial IgG antibodies are widely used for epidemiological and diagnostic purposes [26,35,36]. However, the antifilarial IgG antibody assay provides a relatively low specificity [37]. Cross-reaction with closely related helminths, such as *Strongyloides* spp., has been reported [38]. Specific IgG subclass antibodies have been used to improve the specificity over the total IgG antibodies for the diagnosis of parasitic infections, such as IgG1 and IgG2 for gnathostomiasis [39], IgG2 and IgG4 for echinococcosis [40], and IgG4 for ascariasis [41]. The specific antifilarial IgG4 antibody assay also enhances the sensitivity of the immunodiagnosis for lymphatic filariasis [26]. Detection of antifilarial IgG4 antibodies is associated with the circulating filarial antigen status [35]. The specific antifilarial IgG4 antibody levels elevate in patients with active *W. bancrofti* infection and decline following the DEC treatment [42,43]. However, about 10% of microfilaremia patients were negative for antifilarial IgG4 ELISA [44]. Because antifilarial IgG4 response is quite specific and correlates with the active infection, it is useful for the diagnosis of brugian filariasis.

Antibody detection assays generally use crude filarial extract antigens derived from whole parasites. The major problems are the limited sources of native filarial parasites and common antigens in parasites causing the cross-reactivity. Until now, there is no animal model for *W. bancrofti*, while the animal model for *B. malayi* is limited only in some laboratories. Antibody detection assays using recombinant antigens should be the method of choice for the diagnosis of filariasis, especially for brugian filariasis. Specific antibody detection assays using recombinant antigens have been developed for the diagnosis of lymphatic filariasis. The detection of specific IgG4 antibodies against WbSXP1 or BmSXP1, Bm14 in blood and urine samples shows high sensitivity (85%−100%) and high specificity (99%−100%) [38,45−49]. Although the detection of anti-BmR1 IgG4 antibodies has excellent sensitivity (91%−100%) for the diagnosis of brugian filariasis, it shows only 14%−44% sensitivity for the diagnosis of bancroftian filariasis [46,50−52]. No cross-reaction with other protozoa and helminths has been found [50,51]. Now, assays for anti-BmR1 IgG4 antibodies, such as Brugia rapid test and PanLF rapid test, are commercially available and are selected for use during the transmission assessment survey in the GPELF program by WHO.

The discovery of *Wolbachia* in filarial nematodes raises several intriguing questions regarding the evolution of *Wolbachia* as well as new opportunities for the diagnosis and control of these diseases. An antigen detection system for *Wolbachia*-derived molecules released into the host's circulation after the drug treatment may be useful as posttreatment surveillance for filariasis elimination program. Recent studies on the pathogenesis of filariasis have detected antibodies against *Wolbachia* surface protein (WSP) in rhesus monkeys infected with *B. malayi* [53], in cats infected with *D. immitis* [54], and in both blood and urine from dogs infected with *D. immitis* [55,56]. Subsequently, the specific antibody responses to WSP have been detected in patient sera, and the levels were correlated with the chronic manifestations of lymphatic filariasis [57].

An alternative method for the identification of the active infection is the detection of filarial DNA. DNA-based techniques have been developed to identify and differentiate filarial parasites in humans, animal reservoir hosts, and mosquito vectors. These include DNA hybridization, polymerase chain reaction (PCR) amplification, and quantitative PCR using specific primers. The common specific targets for *W. bancrofti* are *Ssp* I repeat, pWb12 repeat, pWb-35 repeat, and LDR repeat, while common specific targets for *Brugia malayi* are *Hha* I repeat, glutathione peroxidase gene, and mitochondrial DNA [58,59]. However, the sensitivity of these DNA-based techniques is lower than CFA assays or just equivalent to the microscopic methods [24]. Other DNA-based techniques are also developed to increase the sensitivity and differentiate the species of the parasites in one reaction such as multiplex-PCR, PCR-restriction fragment length polymorphism (PCR-RFLP), as well as PCR enzyme-linked immunosorbent assay (PCR-ELISA) [58,59]. Furthermore, because bancroftian filariasis is endemic on the Thai-Myanmar border, the potential now exists for a reemergence of bancroftian filariasis in Thailand. Random amplified polymorphic DNA (RAPD) analysis has proved effective to differentiate Thai and Myanmar strains of *W. bancrofti* [60]. The summary of advantages and disadvantages for the current diagnosis of lymphatic filariasis, including Mf detection, antigen tests, antibody tests, and DNA-based assays, is shown in Table 4.1.

Noninvasive imaging techniques, such as ultrasonography, allow visualization of adult parasites in the lymphatic system, especially in bancroftian filariasis patients

Table 4.1 Summary of advantages and disadvantages for current diagnosis of lymphatic filariasis.

Methods	Available techniques	Advantages	Disadvantages
Detections of microfilariae	• Giemsa's staining • Knott's technique • Membrane filtration	• High specificity • Inexpensive • Easy to perform • Indicate the potential for transmission	• Low sensitivity • Expertize skills • Timing of blood collection because of periodicity • Not correlate with adult worm burden in lymphatics • Amicrofilariamics (i.e., immature females, single-sex infections, effete females, Mf clearing by host's immune response)
Detections of CFA	• ICT • ELISA	• Active infection • Blood collection at any time • High sensitivity • High specificity • Easy to perform (ICT) • Rapid assay (ICT) • Quantitative test (ELISA)	• Expensive • Requirement of two antibodies (monoclonal and polyclonal antibodies) • Only for W. bancrofti • False negative (female dead, infection with males) • Read at exactly time to avoid false positive (ICT) • Time consuming (ELISA) • Laborious steps (ELISA) • Facilities and instruments (ELISA) • Expertize skills (ELISA)
Detections of specific antibodies	• ELISA	• Blood collection at any time • High sensitivity • Quantitative test	• Cross-reactivity • Past infection • Expensive • Time consuming • Laborious steps • Facilities and instruments • Expertize skills
Detections of genetic materials	• PCR • qPCR • RFLP • RAPD	• High sensitivity • High specificity • Differentiation between species and/or strains of the parasites	• Not equate to infection • Time consuming • Laborious steps • Facilities and instruments • Expertize skills

with hydroceles [61−63]. Up to 80% of the infected men show filarial movement termed "filarial dance sign" (FDS) in the scrotal area [62]. Patients with no FDS have a significantly lower microfilarial density than among patients with FDS [62]. Although a patient with testicular discomfort has negative results for filarial microscopy and serology, FDS has been detected by ultrasound [63]. However, ultrasonography could not detect living adult worms in deep tissues, such as in lymphatic vessels and lymph nodes. This tool is not only useful for the diagnosis of the disease but also for the evaluation of the macrofilaricidal activity of treatments [64].

In summary, although alternative diagnostic approaches, such as antibody detection and PCR, have also been considered, antigen detection remains the most valuable tool for the endemicity mapping and certification of elimination necessary to the GPELF. Indeed, the detection of antifilarial antibodies in the serum cannot distinguish between active and past infections. This approach also has low sensitivity due to the cross-reaction of the antigens used in the assays with the antibodies raised by the patient against other parasites. PCR also shows low sensitivity as the filarial DNA must be first extracted from the microfilariae in the blood. As the adult worms reside in the lymphatic system, low levels of filarial DNA can be expected to be found in the serum of amicrofilaremic patients.

4.3 Epidemiology and risk factors

4.3.1 Epidemiology

Lymphatic filariasis is a global health problem and considered to be one of the socioeconomic burden diseases. The disease often occurs in rural areas in the tropics and subtropics. Among the pathogens causing lymphatic filariasis, *W. bancrofti* has the widest geographical distribution worldwide and is prevalent in tropical areas, including Africa, Asia, the Caribbean, Latin America, and many islands of the western and South Pacific Ocean. *B. malayi* is more restricted, being found in southwest India, China, Indonesia, Malaysia, Korea, the Philippines, Vietnam, and Thailand. *B. timori* is restricted in Timor, Flores, Alor, Roti, and southeast Indonesia (Table 4.2) [65].

Table 4.2 Lymphatic filarial parasites in human and their distribution.

Species	Percentage	Distribution	Mosquito vectors
Wuchereria bancrofti	90	Africa, Asia, the Caribbean, Latin America, Islands of the western and south Pacific Ocean	*Culex, Anopheles, Aedes*
Brugia malayi	10	Southwest India, China, Indonesia, Malaysia, Korea, the Philippines, Vietnams, Thailand	*Mansonia, Anopheles, Aedes*
Brugia timori	0.67	Timor, Flores, Alor and Pantar, Roti, southeast Indonesia	*Anopheles*

At the start of GPELF in 2000, 1333 million people in 83 countries in the tropics and subtropics worldwide were at risk (Table 4.3) [66], over 120 million people were infected with about 40 million were seriously debilitated and disfigured by the disease (25 million men with hydrocele and over 15 million people with lymphedema or elephantiasis). The largest number of infected people are present in India (about 45 million) and Sub-Saharan Africa (about 40 million). Although the low prevalence is reported in Egypt and Latin America, some regions in the Pacific Islands show very high prevalence, including Papua New Guinea (72%) and Republic of Tonga (48%) [65].

Since 2000, a cumulative total of 6.7 billion treatments has been distributed to more than 850 million people. Until now, 14 countries (Cambodia, The Cook Islands, Egypt, Maldives, Marshall Islands, Niue, Palau, Sri Lanka, Thailand, Togo, Tonga, Vanuatu, Vietnam, and Wallis and Fortuna Islands) are now acknowledged for the complete elimination of lymphatic filariasis as a public health problem (Table 4.3). Another six countries have successfully implemented strategies and already stopped mass drug administration (MDA). They are under surveillance to validate the elimination of the disease. However, 856 million people in 52 countries worldwide remain threatened by LF and require preventive chemotherapy, 6 of the 52 countries had not started MDA, including Eritrea, Equatorial Guinea, Gabon, Sao Tome and Principe, South Sudan, and New Caledonia [67].

Even after achieving a low infection rate of the disease, the migration may serve as the risk of transmission, such as Myanmar migrant workers in Thailand. Therefore, continuous monitoring and dispensing treatments to the migrants need to be performed to prevent the reemergence of the disease. In some endemic countries, surveillance for the transmission of lymphatic filariasis after stopping MDA shows the reemergence of *B. malayi* despite the successful elimination of *W. bancrofti* [68].

4.3.2 Risk factors

The risk factors relevant to brugian and bancroftian filariasis are listed in Table 4.4 [69−86]. These factors include gender, occupation, age, parental infection, education status, level of knowledge, use of mosquito nets, length of stay, housing, participation in MDA programs, the proximity of swamp or pooled water, and inadequate clothing. For effective filariasis prevention focusing on risk factors, people who live in the endemic area should protect themselves from the bites of mosquito vectors by wearing permethrin-treated clothing when working outdoors [87], using mosquito spatial repellents [88], entomopathogenic fungi for controlling both mosquito larva and adult mosquitoes [89], and sleeping under mosquito nets treated with nonpyrethroid or with pyrethroid plus nonpyrethroid, plus insect growth regulator, or plus piperonyl butoxide [90]. Moreover, the implementation of education and training to disseminate knowledge regarding filariasis prevention and control, as well as vector control in the communities via the information media and printed materials, and encouraging the people in the community to participate in MDA programs annually (or biannually) should be performed [86,91,92].

Table 4.3 Endemic countries of lymphatic filariasis and status for the elimination as of 2017 [6€].

Africa PRG (39 countries)	American PRG (7 countries)	Eastern Mediterranean PRG (3 countries)	PacCARE PRG (17 countries)	Mekong-Plus PRG (8 countries)	South Asia PRG (9 countries)
Countries that achieve the elimination criteria					
		Egypt	Cook Islands, Marshall Islands, Niue, Palau, Tonga, Vanuatu, Wallis and Futuna Islands	Cambodia, Vietnam	Maldives, Sri Lanka, Thailand
Countries that stopped MDA and under surveillance					
Malawi	Brazil	Yemen	American Samoa, Kiribati		Bangladesh
Countries that still require MDA					
Benin, Burkina Faso, Comoros, Ghana, Kenya, Nigeria, Tanzania, Uganda, Angola, Burundi, Cameroon, Cape Verde, Central African Republic, Chad, Congo, Côte d'Ivoire, Democratic Republic of the Congo, Equatorial Guinea, Ethiopia, Gabon, Gambia, Guinea	Dominican Republic, Costa Rica, Guyana, Haiti, Suriname, Trinidad and Tobago	Sudan	Fiji, French Polynesia, Micronesia, Samoa, Tuvalu, Solomon Islands, New Caledonia, Papua New Guinea	Brunei Darussalam, China, Lao PDR, Korea Malaysia, Philippines	India, Indonesia, Myanmar, Nepal, Timor-Leste

Guinea-Bissau
Liberia
Madagascar
Mali
Mauritius
Mozambique
Niger
Réunion
Rwanda
São Tomé and Principe
Senegal
Seychelles
Sierra Leone
Zambia
Zimbabwe

Table 4.4 Risk factors for human lymphatic filariasis.

Variables	Filaria species	Descriptions
Gender and occupation-dependent exposure to mosquito vectors	*W. bancrofti*	Microfilaremia increases in males aged 15—44 years [69] Males who work outdoors, such as on plantations, have an increased degree of exposure time to mosquito vectors [70]. Multivariate analysis of antigenemia and microfilaremia showed an increased risk for males. Hunting and fishing at night carried increased risk for antigenemia [71]. Outdoor jobs, such as farming, are more vulnerable to being bitten by mosquito vectors than housewives [72].
Age	*W. bancrofti*	Antigenemia and microfilaremia increase with age up to 20 years [71].
Parental infection	*W. bancrofti*	Maternal infection is a risk factor for filarial infection in children and is associated with the altered parasite-specific immune reactivity [73—78]. Paternal infection is a risk factor for children ages 5—14 years [79]. Paternal infections are a significant risk factor for infection in ages 4—10 years [80] and 10—16 years [81].
Educational status	*W. bancrofti* *B. malayi* *W. bancrofti* *B. timori*	Those who had never been to school showed an increase in microfilaremia and elephantiasis [82]. Multivariate analysis of microfilaremia showed an increased risk for nongraduates [83]. Persons with a low level of education showed a significantly higher risk for filarial infection [84].
Level of knowledge	*B. malayi* *W. bancrofti* *B. timori*	Regarding the symptoms, transmission, prevention, and treatment of filariasis, a significant connection was shown between the levels of knowledge of the respondents and the incidence of filariasis [84,85].
Use of mosquito nets	*B. malayi* *W. bancrofti* *B. timori*	Microfilaremia increased for females [69] who do not sleep under a mosquito net [85,86].
Length of stay	*W. bancrofti*	Microfilaremia increases from living for more than 5 years in an area exposed to the mosquito vector [69,79].

Table 4.4 Risk factors for human lymphatic filariasis.—*cont'd*

Variables	Filaria species	Descriptions
Housing	*W. bancrofti*	Microfilaremia is related to living in huts, thatched and tiled houses as compared to houses constructed of reinforced concrete [83].
Participation in the MDA program	*W. bancrofti*	Bivariate and multivariate analyses of microfilaremia show an increased risk for nonparticipants in the MDA program [83].
Proximity to swamp or pooled water	*B. malayi* *W. bancrofti* *B. timori*	There is a relationship between the presence of pooled water near the house and the incidence of filariasis [84—86].
Inadequate clothing	*B. malayi* *W. bancrofti*	There is a higher risk of acquiring filarial infection for those not using long sleeves and pants [85].

4.4 Approaches to control and elimination through MDA

MDA under GPELF established by the WHO was launched in the year 2000, with the goal of successful interruption of filarial transmission, prevention, and relief of disability caused by the disease, in all endemic areas by the year 2020 [7]. The antifilarial drugs include a single dose of DEC (6 mg/kg) with albendazole (ABZ, 400 mg) administered annually in each MDA for at least 4—6 years [93,94]. Recently, WHO has recommended a single dose of a triple-drug treatment for bancroftian filariasis, IVM (200 mcg/kg) in combination with DEC (6 mg/kg) and ABZ (400 mg) [95]. This regimen was significant in clearing blood microfilaria (embryostatic and/or embryocidal effects) for 2 years and reducing filarial antigen levels (partial macrofilaricidal effect) after treatment with no severe adverse effects, as compared to the DEC/ABZ standard [96,97].

The expected effective coverage of multiple rounds in the MDA program in each endemic area (>65% coverage) is recommended by WHO. The information regarding country status in implementing MDA for lymphatic filariasis as of 2017 showed that 51 endemic countries were considered to require MDA. Equatorial Guinea, Gabon, Sao Tome and Principe, South Sudan and New Caledonia had not started MDA. Angola, Cameroon, Chad, Central African Republic, Congo, Democratic Republic of Congo, Ethiopia, Guinea-Bissau, Madagascar, Nigeria, Guyana, Sudan, and Papua New Guinea have implemented MDA but not yet in all endemic implementation units [67]. The major reasons for the persistence of the disease after MDA campaigns include lack of proper knowledge and motivation in the community, patients' fear of the adverse drug reactions (fever, headache, and dizziness), a lack of home visits by health workers during MDA to build trust in the MDA regimen, and budget limitations [98—101].

To achieve the goal of elimination of lymphatic filariasis by the year 2020, the Eco-Health/One Health approach that succeeded in controlling liver fluke in Thailand [102] should be implemented in the areas of endemic lymphatic filariasis. The stakeholders include nongovernmental organizations, teachers, students, private medical practitioners, public health workers, the communities, and religious leaders who should be involved to encourage people in the endemic areas to participate in the MDA program [103,104]. In areas with high microfilaria prevalence, more than six rounds of MDA should be carried out to interrupt disease transmission [92,105,106]. Moreover, additional strategies, such as community-based integrated vector control [70,91,107], improving compliance through strict MDA supervision, door-to-door drug distribution [108], biannual and semiannual treatment in high-incidence endemic areas [92,109], and drug resistance monitoring [92] should be integrated into the MDA program to maintain and sustain a lymphatic filariasis-free country.

4.5 Next-generation vaccine, drugs, and diagnostics

4.5.1 Next-generation vaccines

The prophylaxis of lymphatic filariasis using vaccination is still in demand. The subcutaneous route of immunization with live irradiation-attenuated infective stage larvae (L3) *B. malayi* showed protective immunity and a 91% decrease in worm burden after challenge in the infected-jird model [110]. To enhance immunogenicity, the recombinant abundant larval transcript-2 protein (ALT-2), secreted mainly by L3, fused with the Tuftsin immunopotentiator and produced by the *Pichia partoris* expression system (P-TULF-ALT-2), was studied. A significantly increased level of IgG2b isotype in BALB/c mice and an increase in IL2, IL-5, and IFN-γ cytokines in the culture supernatant were observed [111]. A DNA vaccine with multiple (cocktail) antigens, such as the potential vaccine candidate antigen *B. malayi* ALT-2 (BmALT-2) and thioredoxin peroxidase (BmTPX), induced protective immunity (78% cytotoxicity) caused by a high IgG2a isotype titer and an increase in IFN-γ production in the BALB/c mouse model, which represented a Th1-type response [112]. In addition, a study in nonhuman primate (rhesus macaques) using r*Bm*HAXT plus AL019 (alum plus glucopyranosyl lipid adjuvant-stable emulsion) showed 57% protection against infection with IgG1, IgG2, and IgG3 production. Further investigation should be studied in humans as this vaccine formulation provides over 50% protection against the worm burden, as advised in the WHO's recommendation [113].

Recently, the discovery of exosome-like extracellular vesicles (EVs) containing large amounts of substances composed of proteins, lipids, DNA, RNA, and noncoding RNA (microRNA) was reported in various helminths. EVs secreted by the helminths included *B. malayi*, which is involved in helminth pathogenesis, controls parasite survival, and modulates host immune response [114−116]. A previous study demonstrated that EV-alum combination induced high titers of IgM, IgG1, and IgA in mouse model, followed by macrophage internalization, and protected the mice from *Heligmosomoides polygyrus* infection [117]. The host–parasite interaction involving EVs could provide new strategies to prepare synthetic vesicle-mediated vaccine delivery systems [118].

4.5.2 Next-generation drugs

The limitations of doxycycline, a macrofilaricidal regimen targeting *Wolbachia* endosymbiont, in a long-time course treatment (daily administration for 4–6 weeks) are its contraindication in children (<8 years) and pregnant women [119,120]. This drug at a high concentration of 32 mcg/mL could inhibit *Wolbachia* growth by 50% (IC50) in *B. malayi* mf at 12 hours *in vitro* [121]. To overcome this disadvantage of the time frame and concentration of treatment, the registered regimens were studied for the combination, derivative substitution, dose, and time adjustment in the *B. malayi* murine infection model. The study demonstrated that rifampicin (RIF) combined with ABZ showed a synergizing effect of >99% *Wolbachia* reduction at 7 days after administration [122]. Minocycline, a tetracycline derivative, was able to reduce *Wolbachia* more powerfully (1.7-fold) as compared to doxycycline, after 28 days [123]. In addition, high doses of RIF (35 mg/kg) for 7 days resulted in >90% *Wolbachia* depletion [124]. Further study regarding minocycline or RIF combined with the registered regimens, such as anti-*Wolbachia* and antifilarial drugs, should be investigated to reduce the duration of treatment and adverse drug effects.

To date, the Anti-*Wolbachia* Consortium combined with high-throughput anti-*Wolbachia* whole-cell screen was initiated resulting in the screening of 25,000 compounds per month, in selected suitable macrofilaricidal drugs for lymphatic filariasis and onchocerciasis, tested with 1.3 million AstraZeneca's compound library [125–130]. Thus around 1.3 million compounds were successfully screened using *Wolbachia* cell-based C6/36 (*w*AlbB) cells (30 μM–1.5 nM final concentration, *B. malayi* microfilaria; 5 μM final concentration) and using in silico analysis [130]. The top five chemotypes (1A–5A) obtained from screening showed >70% *Wolbachia* depletion after 2 days of exposure, and thus proved to be superior to doxycycline in the same assay [130].

Several macrofilaricidal agents and new drugs have been recommended including flubendazole (FBZ), aptamers, and auranofin with identification of the new drug targets: thioredoxin reductase, glutathione-S-transferases, and Ca^{2+} binding protein and calumenin.

Flubendazole, a registered benzimidazole anthelmintic, causes the effect on embryogenesis by decreasing mf release [131] and downregulating the genes encoding cuticle components of adult female *B. malayi in vitro* [132]. However, oral amorphous solid dispersion (ASD) formulation of FBZ demonstrated no significant microfilaricidal effect on *B. malayi* mf in the rodent model [133] and did not reduce total adult *B. pahangi* burden in the infected jird model [134]. Future studies should evaluate female worm viability and sterilization with ASD FBZ or other FBZ oral forms prepared by amorphous technology to enhance the oral bioavailability of FBZ [135].

The discovery of aptamers, short DNA/RNA oligonucleotides that can bind specific targets, selected by the SELEX (Systematic Evolution of Ligands by EXponential enrichment) process [136], may be an alternative method for filarial diagnosis, prevention, and treatment. The studies involved in parasite-specific aptamers reported, for example, *Schistosomiasis japonicum*' eggs detection in stool and liver samples, using LC-6 and LC-15 aptamers that bind to the *S. japonicum* egg surface [137]. Various aptamers can inhibit the survival and infection of parasites, such as *Trypanosoma cruzi in vitro* invasion, by targeting surface components of the parasite: HS6 (target heparin sulfate), F4 (target fibronectin), L28 (target laminin), and T8 (target thrombospondin) [138]. Aptamers were also used as a new tool for drug delivery by its internalization through the receptors [139—142]. Further studies can explore the design of new aptamers that can serve as a potential new drug to block *Wolbachia* and filarial parasite longevity.

Various antifilarial drug targets such as enzymes involved in homeostasis and detoxification have been identified. Auranofin is an oral gold-containing antirheumatic, antiprotozoa and anticancer regimen, an FDA approved drug targeting the thioredoxin reductase, an enzyme that maintains the intracellular redox homeostasis [143,144]. This drug has a macrofilaricidal effect on the adult *B. pahangi and B. malayi in vitro* when incubated for 4 days and on adult *B. pahangi* in the jird model when exposed for 28 days [145]. Glutathione-S-transferases (GST), the major detoxification enzyme, is responsible for the cellular redox condition of many helminths [146]. *W. bancrofti* and *B. malayi* GST protein shared 98% identity. Using an in silico virtual screening approach, the curcumin, which is the natural compound that can inhibit both BmGST and WbGST, is proposed as a future candidate for an alternative drug for not only filarial parasites but also helminth infections [147]. Moreover, the Ca^{2+} binding protein, calumenin, responsible for cuticle development and fertility in *B. malayi*, was identified as a candidate drug target. In silico screening found that itraconazole, a triazole antifungal drug, is a potential calumenin-targeting ligand, when studied in a *Caenorhabditis elegans* mutant model [148].

The biosynthesis pathways of filarial and *Wolbachia* have been investigated for the drug targets. The *Wolbachia* lysine biosynthetic pathway was proposed as an attractive novel target for antifilarial regimens. Aspartate semialdehyde dehydrogenase (ASDase), which is absent in humans, is one of the nine enzymes responsible for lysine biosynthesis. The potent lead molecules that can cause the enzyme dysfunctional by forming the stable complex with ASDase will be further analyzed using computational analysis for their development as potential lead compounds [149]. Recently, the presence of iron transporter, divalent metal transporter 1 DMT1, in all stages of *B. malayi* especially adult female has also gathered significant scientific attention as the drug target [150]. Basically, the iron provided by *Wolbachia* is necessary for embryos and uterine microfilariae development, and this cannot be synthesized by the parasite on its own [151]. The heme biosynthesis pathway, as well as DMT1, is, therefore, an attractive antifilarial drug target. In addition, the type IV secretion system (T4SS) and riboflavin (vitamin B2) biosynthesis in

Wolbachia of *B. malayi* are also considered as potential drug targets. *B. malayi* requires the supplementation of riboflavin from *Wolbachia* and previous studies have demonstrated that two identified transcription factors, *w* BmxR1 and *w* BmxR2, coregulate transcription pathways [152]. Drug regimens targeting *Wolbachia* as well as the transcription factors and various putative T4SS effector proteins [153] can be investigated for disrupting the mutualistic symbiosis.

4.5.3 Next-generation diagnostics

Traditionally, the diagnosis of active infection of lymphatic filariasis depends on the detection of microfilariae in blood circulation. The limitations of this method are a low sensitivity and nocturnal periodicity of the parasites. The detection of CFA is used as the gold standard for the diagnosis. However, CFA detection is now available only for bancroftian filariasis. Therefore, the development of diagnostic techniques that can diagnose both brugian and bancroftian filariasis is still needed.

The **shotgun metagenomic assay** is a technique for sequencing and analyzing vast amounts of DNA directly extracted from clinical samples without amplification of specific targets or isolation of pathogens. It is the unbiased approach to identify pan-pathogens and difficult-to-diagnose pathogens missed by targeted sequencing and culture methods. This method has been widely used in the detection of emerging pathogens, de novo pathogen identification, drug resistance gene detections, as well as viral and bacterial outbreak investigations [154,155]. Recently, the shotgun metagenomic assay has been successfully applied to diagnose a nonendemic patient with atypical brugian filariasis who failed to be diagnosed by detection of Mf, serology, as well as PCR-based methods for several gene targets [156]. However, this assay requires expert skills and special instruments that might be not available in endemic areas of lymphatic filariasis. Improvement in technologies and bioinformatics software tools leads to a decrease in the cost and allows widespread use of the assay in the future.

Aptamers, small synthetic single-stranded DNA or RNA molecules, have high flexibility and can change into many forms. They form high-affinity noncovalent bond with several specific molecules, including nucleic acids, proteins, small organic compounds, antibiotics, as well as toxic molecules. These aptamers can be chemically synthesized with no batch-to-batch variation with unlimited shelf life. Because of the small size, aptamers show less immunogenicity than antibodies. From these advantages, the aptamers seem to be better than antibodies and could replace antibodies in the future. Aptamers have been applied for diagnostic and therapeutic use for decades [157].

For diagnostic use, **aptamers** are used as a recognizing agent for the targets in several format tests with higher sensitivity than antibody-based assays, such as aptamer-linked immobilized sorbent assay (ALISA) for the identification of enterotoxigenic *Staphylococcus aureus* from food samples [158], enzyme-linked oligonucleotide assay (ELONA) for the diagnosis of *Leishmania infantum* [159,160]. Not only in the enzyme method, but aptamers are also applied to use in electrochemical methods. Aptamers are coated on several types of electrode surfaces. Binding with a specific target leads to conformational changing of aptamers and then results in changing of electric current that is detected by the detectors [161]. These methods are very sensitive but require expert skills and special instruments.

The simplest method using aptamers as the recognizing agent is a **colorimetric-based optical sensor.** Aptamers form a noncovalent bond with gold-nanoparticles (GNPs). Binding with specific targets leads to color change that can be observed by naked eyes. This colorimetric-based optical sensor is also applied as a DNA-based aptamer lateral flow method in several designs [162]. For example, the control line contains ssDNA specific to the variable region of aptamers, while the test line contains ssDNA specific to a linker sequence. In the absence of targets, aptamers coated on GNPs flow through the membrane surface and bind to ssDNA on the control line and the test line with equal affinity resulting in the equal intensity of two red lines. In contrast, in the presence of targets binding to a variable region of aptamers, there is no free variable region on the aptamers. Therefore, no binding of aptamers to the control line results in faint red color on the control line, while aptamer-target complex binding with ssDNA on test line through linker sequence results in the strong red color on the test line. Aptamer lateral flow assays are developed to detect small molecules such as ampicillin detection [163], cholera toxin [164]. Unfortunately, aptamer-based biosensors have never been developed for the diagnosis of lymphatic filariasis. The development of aptamer-based biosensors for diagnosis of filariasis might be useful not only for diagnosis but also for monitoring and control of the disease.

4.6 Concluding remarks

Lymphatic filariasis elimination is not only the target of integrative approaches such as MDA program participation, vector control, and knowledge implementation but also the investigation of novel potential regimens. The major aim is that the 7-day treatment duration should be safe for children under 8 years old and pregnant women. The essential pathways relevant to the filarial—*Wolbachia* relationship can be further investigated for the identification of new effective drug regimens, with the potential for interrupting the mutualistic symbiosis. Next-generation vaccine focusing on the effective adjuvant, nanodelivery system of DNA vaccine and EVs as a method of blocking the host—parasite interaction may be used for control and disease prevention. Moreover, as a large number of people in the world are infected with helminth endoparasites such as lymphatic filariasis, soil-transmitted helminths, and trematodes, the potential conservation of antigens Sm-14, subolesin/akarin, and P0 can also be studied for the development of a pan-parasitic vaccine [165].

References

[1] Fischer P, Supali T, Maizels RM. Lymphatic filariasis and *Brugia timori*: prospects for elimination. Trends in Parasitology August 2004;20(8):351−5.

[2] Ottesen EA, Duke BO, Karam M, Behbehani K. Strategies and tools for the control/elimination of lymphatic filariasis. Bulletin of the World Health Organization 1997;75(6):491−503.

[3] World Health Organization. Lymphatic filariasis: the disease and its control. Fifth report of the WHO Expert Committee on Filariasis. World Health Organisation Technical Report Series 1992;821:1−71.

[4] Melrose WD. Lymphatic filariasis: new insights into an old disease. International Journal for Parasitology July 2002;32(8):947−60.

[5] Behbehani K. Candidate parasitic diseases. Bulletin of the World Health Organization 1998;76(Suppl. 2):64−7.

[6] World Health Organization. reportWorld health report 2004. Geneva: World Health Organization.

[7] Yamey G. Global alliance launches plan to eliminate lymphatic filariasis. The British Medical Journal January 29, 2000;320(7230):269.

[8] Vanamail P, Subramanian S, Das PK, Pani SP, Rajagopalan PK. Estimation of fecundic life span of *Wuchereria bancrofti* from longitudinal study of human infection in an endemic area of Pondicherry (south India). Indian Journal of Medical Research July 1990;91:293−7.

[9] Witt C, Ottesen EA. Lymphatic filariasis: an infection of childhood. Tropical Medicine and International Health August 2001;6(8):582−606.

[10] Ottesen EA. The Wellcome Trust Lecture. Infection and disease in lymphatic filariasis: an immunological perspective. Parasitology 1992;104(Suppl. 1):S71−9.

[11] Figueredo-Silva J, Noroes J, Cedenho A, Dreyer G. The histopathology of bancroftian filariasis revisited: the role of the adult worm in the lymphatic-vessel disease. Annals of Tropical Medicine and Parasitology September 2002;96(6):531−41.

[12] Tsanglao WR, Nandan D, Chandelia S, Arya NK, Sharma A. Filarial tropical pulmonary eosinophilia: a condition masquerading asthma, a series of 12 cases. Journal of Asthma July 2019;56(7):791−8.

[13] Chhotray GP, Mohapatra M, Acharya AS, Ranjit MR. A clinico-epidemiological perspective of lymphatic filariasis in Satyabadi block of Puri district, Orissa. Indian Journal of Medical Research August 2001;114:65−71.

[14] Brabin L. Sex differentials in susceptibility to lymphatic filariasis and implications for maternal child immunity. Epidemiology and Infection October 1990;105(2):335−53.

[15] McLaren DJ, Worms MJ, Laurence BR, Simpson MG. Micro-organisms in filarial larvae (Nematoda). Transactions of the Royal Society of Tropical Medicine and Hygiene 1975;69(5−6):509−14.

[16] Werren JH. Biology of *Wolbachia*. Annual Review of Entomology 1997;42:587−609.

[17] Bandi C, Trees AJ, Brattig NW. *Wolbachia* in filarial nematodes: evolutionary aspects and implications for the pathogenesis and treatment of filarial diseases. Veterinary Parasitology July 12, 2001;98(1−3):215−38.

[18] Sungpradit S, Nuchprayoon S. *Wolbachia* of arthropods and filarial nematodes: biology and applications. Chulalongkorn Medical Journal November−December 2010;54(6):605−21.

[19] Eberhard ML, Roberts IM, Lammie PJ, Lowrie Jr RC. Comparative densities of *Wuchereria bancrofti* microfilaria in paired samples of capillary and venous blood. Tropical Medicine and Parasitology December 1988;39(4):295−8.

[20] Michael E, Grenfell BT, Bundy DA. The association between microfilaraemia and disease in lymphatic filariasis. Proceedings of the Royal Society B: Biological Sciences April 22, 1994;256(1345):33−40.

[21] Weil GJ, Liftis F. Identification and partial characterization of a parasite antigen in sera from humans infected with *Wuchereria bancrofti*. The Journal of Immunology May 1, 1987;138(9):3035−41.

[22] More SJ, Copeman DB. A highly specific and sensitive monoclonal antibody-based ELISA for the detection of circulating antigen in bancroftian filariasis. Tropical Medicine and Parasitology December 1990;41(4):403−6.

[23] Rocha A, Addiss D, Ribeiro ME, Norões J, Baliza M, Medeiros Z, et al. Evaluation of the Og4C3 ELISA in *Wuchereria bancrofti* infection: infected persons with undetectable or ultra-low microfilarial densities. Tropical Medicine and International Health December 1996;1(6):859−64.

[24] Nuchprayoon S, Yentakam S, Sangprakarn S, Junpee A. Endemic bancroftian filariasis in Thailand: detection by Og4C3 antigen capture ELISA and the polymerase chain reaction. Medical Journal of the Medical Association of Thailand September 2001; 84(9):1300−7.

[25] Nuchprayoon S, Porksakorn C, Junpee A, Sanprasert V, Poovorawan Y. Comparative assessment of an Og4C3 ELISA and an ICT filariasis test: a study of Myanmar migrants in Thailand. Asian Pacific Journal of Allergy and Immunology December 2003;21(4):253−7.

[26] Nuchprayoon S, Sanprasert V, Porksakorn C, Nuchprayoon I. Prevalence of bancroftian filariasis on the Thai-Myanmar border. Asian Pacific Journal of Allergy and Immunology September 2003;21(3):179−88.

[27] Oliveira P, Braga C, Alexander N, Brandão E, Silva A, Wanderley L, et al. Evaluation of diagnostic tests for *Wuchereria bancrofti* infection in Brazilian schoolchildren. Revista da Sociedade Brasileira de Medicina Tropical 2014 May−June;47(3): 359−66.

[28] Weil GJ, Jain DC, Santhanam S, Malhotra A, Kumar H, Sethumadhavan KV, et al. A monoclonal antibody-based enzyme immunoassay for detecting parasite antigenemia in bancroftian filariasis. The Journal of Infectious Diseases August 1987; 156(2):350−5.

[29] Weil GJ, Ramzy RM, Chandrashekar R, Gad AM, Lowrie Jr RC, Faris R. Parasite antigenemia without microfilaremia in bancroftian filariasis. The American Journal of Tropical Medicine and Hygiene September 1996;55(3):333−7.

[30] Wanji S, Amvongo-Adjia N, Njouendou AJ, Kengne-Ouafo JA, Ndongmo WP, Fombad FF, et al. Further evidence of the cross-reactivity of the Binax NOW® Filariasis ICT cards to non-*Wuchereria bancrofti* filariae: experimental studies with *Loa loa* and *Onchocerca ochengi*. Parasites and Vectors May 5, 2016;9:267.

[31] Hertz MI, Nana-Djeunga H, Kamgno J, Jelil Njouendou A, Chawa Chunda V, Wanji S, et al. Identification and characterization of *Loa loa* antigens responsible for cross-reactivity with rapid diagnostic tests for lymphatic filariasis. PLoS Neglected Tropical Diseases November 16, 2018;12(11):e0006963.

[32] Gounoue-Kamkumo R, Nana-Djeunga HC, Bopda J, Akame J, Tarini A, Kamgno J. Loss of sensitivity of immunochromatographic test (ICT) for lymphatic filariasis

diagnosis in low prevalence settings: consequence in the monitoring and evaluation procedures. BMC Infectious Diseases December 23, 2015;15:579.

[33] Weil GJ, Curtis KC, Fakoli L, Fischer K, Gankpala L, Lammie PJ, et al. Laboratory and field evaluation of a new rapid test for detecting *Wuchereria bancrofti* antigen in human blood. The American Journal of Tropical Medicine and Hygiene July 2013;89(1):11−5.

[34] Figueredo-Silva J, Jungmann P, Norões J, Piessens WF, Coutinho A, Brito C, et al. Histological evidence for adulticidal effect of low doses of diethylcarbamazine in bancroftian filariasis. Transactions of the Royal Society of Tropical Medicine and Hygiene 1996 March−April;90(2):192−4.

[35] Kwan-Lim GE, Forsyth KP, Maizels RM. Filarial-specific IgG4 response correlates with active *Wuchereria bancrofti* infection. The Journal of Immunology December 15, 1990;145(12):4298−305.

[36] Terhell AJ, Haarbrink M, van den Biggelaar A, Mangali A, Sartono E, Yazdanbakhsh M. Long-term follow-up of treatment with diethylcarbamazine on anti-filarial IgG4: dosage, compliance, and differential patterns in adults and children. The American Journal of Tropical Medicine and Hygiene January 2003; 68(1):33−9.

[37] Chanteau S, Glaziou P, Moulia-Pelat JP, Plichart C, Luquiaud P, Cartel JL. Low positive predictive value of anti-*Brugia malayi* IgG and IgG4 serology for the diagnosis of *Wuchereria bancrofti*. Transactions of the Royal Society of Tropical Medicine and Hygiene 1994 November−December;88(6):661−2.

[38] Muck AE, Pires ML, Lammie PJ. Influence of infection with non-filarial helminths on the specificity of serological assays for antifilarial immunoglobulin G4. Transactions of the Royal Society of Tropical Medicine and Hygiene 2003 January−February; 97(1):88−90.

[39] Nuchprayoon S, Sanprasert V, Suntravat M, Kraivichian K, Saksirisampant W, Nuchprayoon I. Study of specific IgG subclass antibodies for diagnosis of *Gnathostoma spinigerum*. Parasitology Research 2003 September;91(2):137−43.

[40] Khabiri AR, Bagheri F, Assmar M, Siavashi MR. Analysis of specific IgE and IgG subclass antibodies for diagnosis of *Echinococcus granulosus*. Parasite Immunology August 2006;28(8):357−62.

[41] Chatterjee BP, Santra A, Karmakar PR, Mazumder DN. Evaluation of IgG4 response in ascariasis by ELISA for serodiagnosis. Tropical Medicine and International Health October 1996;1(5):633−9.

[42] Atmadja AK, Atkinson R, Sartono E, Partono F, Yazdanbakhsh M, Maizels RM. Differential decline in filaria-specific IgG1, IgG4, and IgE antibodies in *Brugia malayi*-infected patients after diethylcarbamazine chemotherapy. The Journal of Infectious Diseases December 1995;172(6):1567−72.

[43] Wamae CN, Njenga SM, Ngugi BM, Mbui J, Njaanake HK. Evaluation of effectiveness of diethylcarbamazine/albendazole combination in reduction of *Wuchereria bancrofti* infection using multiple infection parameters. Acta Tropica September 2011; 120(Suppl. 1):S33−8.

[44] Marley SE, Lammie PJ, Eberhard ML, Hightower AW. Reduced antifilarial IgG4 responsiveness in a subpopulation of microfilaremic persons. The Journal of Infectious Diseases December 1995;172(6):1630−3.

[45] Rao KV, Eswaran M, Ravi V, Gnanasekhar B, Narayanan RB, Kaliraj P, et al. The *Wuchereria bancrofti* orthologue of *Brugia malayi* SXP1 and the diagnosis of

bancroftian filariasis. Molecular and Biochemical Parasitology March 15, 2000; 107(1):71–80.

[46] Abdul Rahman R, Hwen-Yee C, Noordin R. Pan LF-ELISA using BmR1 and BmSXP recombinant antigens for detection of lymphatic filariasis. Filaria Journal October 26, 2007;6:10.

[47] Noordin R, Itoh M, Kimura E, Abdul Rahman R, Ravindran B, Mahmud R, et al. Multicentre evaluations of two new rapid IgG4 tests (WB rapid and panLF rapid) for detection of lymphatic filariasis. Filaria Journal October 26, 2007;6:9.

[48] Samad MS, Itoh M, Moji K, Hossain M, Mondal D, Alam MS, et al. Enzyme-linked immunosorbent assay for the diagnosis of *Wuchereria bancrofti* infection using urine samples and its application in Bangladesh. Parasitology International December 2013; 62(6):564–7.

[49] Rahman MA, Yahathugoda TC, Tojo B, Premaratne P, Nagaoka F, Takagi H, et al. A surveillance system for lymphatic filariasis after its elimination in Sri Lanka. Parasitology International February 2019;68(1):73–8.

[50] Rahmah N, Shenoy RK, Nutman TB, Weiss N, Gilmour K, Maizels RM, et al. Multicentre laboratory evaluation of Brugia Rapid dipstick test for detection of brugian filariasis. Tropical Medicine and International Health October 2003;8(10):895–900.

[51] Lammie PJ, Weil G, Noordin R, Kaliraj P, Steel C, Goodman D, et al. Recombinant antigen-based antibody assays for the diagnosis and surveillance of lymphatic filariasis - a multicenter trial. Filaria Journal 2004;3(1):9.

[52] Noordin R, Wahyuni S, Mangali A, Huat LB, Yazdanbakhsh M, Sartono E. Comparison of IgG4 assays using whole parasite extract and BmR1 recombinant antigen in determining antibody prevalence in brugian filariasis. Filaria Journal August 12, 2004;3(1):8.

[53] Punkosdy GA, Dennis VA, Lasater BL, Tzertzinis G, Foster JM, Lammie PJ. Detection of serum IgG antibodies specific for *Wolbachia* surface protein in rhesus monkeys infected with *Brugia malayi*. The Journal of Infectious Diseases August 1, 2001; 184(3):385–9.

[54] Bazzocchi C, Ceciliani F, McCall JW, Ricci I, Genchi C, Bandi C. Antigenic role of the endosymbionts of filarial nematodes: IgG response against the *Wolbachia* surface protein in cats infected with *Dirofilaria immitis*. Proceedings of the Royal Society B: Biological Sciences December 22, 2000;267(1461):2511–6.

[55] Morchón R, Carretón E, Grandi G, González-Miguel J, Montoya-Alonso JA, Simón F, et al. Anti-*Wolbachia* Surface Protein antibodies are present in the urine of dogs naturally infected with *Dirofilaria immitis* with circulating microfilariae but not in dogs with occult infections. Vector Borne and Zoonotic Diseases January 2012;12(1): 17–20.

[56] Ciuca L, Simòn F, Rinaldi L, Kramer L, Genchi M, Cringoli G, et al. Seroepidemiological survey of human exposure to *Dirofilaria* spp. in Romania and Moldova. Acta Tropica November 2018;187:169–74.

[57] Punkosdy GA, Addiss DG, Lammie PJ. Characterization of antibody responses to *Wolbachia* surface protein in humans with lymphatic filariasis. Infection and Immunity 2003;71(9):5104–14.

[58] Nuchprayoon S, Junpee A, Poovorawan Y, Scott AL. Detection and differentiation of filarial parasites by universal primers and polymerase chain reaction-restriction fragment length polymorphism analysis. The American Journal of Tropical Medicine and Hygiene November 2005;73(5):895–900.

[59] Nuchprayoon S. DNA-based diagnosis of lymphatic filariasis. Southeast Asian Journal of Tropical Medicine and Public Health September 2009;40(5):904—13.

[60] Nuchprayoon S, Junpee A, Poovorawan Y. Random amplified polymorphic DNA (RAPD) for differentiation between Thai and Myanmar strains of *Wuchereria bancrofti*. Filaria Journal July 30, 2007;6:6.

[61] Amaral F, Dreyer G, Figueredo-Silva J, Noroes J, Cavalcanti A, Samico SC, et al. Live adult worms detected by ultrasonography in human bancroftian filariasis. The American Journal of Tropical Medicine and Hygiene June 1994;50(6):753—7.

[62] Norões J, Addiss D, Amaral F, Coutinho A, Medeiros Z, Dreyer G. Occurrence of living adult *Wuchereria bancrofti* in the scrotal area of men with microfilaraemia. Transactions of the Royal Society of Tropical Medicine and Hygiene 1996 January—February;90(1):55—6.

[63] Wiggers JB, Jang HJ, Keystone JS. Case report: filaria or megasperm? A cause of an ultrasonographic "filarial dance sign". The American Journal of Tropical Medicine and Hygiene July 2018;99(1):102—3.

[64] Marriott AE, Sjoberg H, Tyrer H, Gamble J, Murphy E, Archer J, et al. Validation of ultrasound bioimaging to predict worm burden and treatment efficacy in preclinical filariasis drug screening models. Scientific Reports April 12, 2018;8(1):5910.

[65] Michael E, Bundy DA, Grenfell BT. Re-assessing the global prevalence and distribution of lymphatic filariasis. Parasitology April 1996;112(Pt 4):409—28.

[66] World Health Organization. Global programme to eliminate lymphatic filariasis: progress report, 2016. Weekly Epidemiological Record October 6, 2017;92(40):594—607.

[67] World Health Organization. Global programme to eliminate lymphatic filariasis: progress report, 2017. Weekly Epidemiological Record November 2, 2018;91(44):589—604.

[68] Chandrasena NT, Premaratna R, Samarasekera DS, de Silva NR. Surveillance for transmission of lymphatic filariasis in Colombo and Gampaha districts of Sri Lanka following mass drug administration. Transactions of the Royal Society of Tropical Medicine and Hygiene December 2016;110(10):620—2.

[69] de Albuquerque Mde F, Marzochi MC, Ximenes RA, Braga MC, Silva MC, Furtado AF. Bancroftian filariasis in two urban areas of Recife, Brasil: the role of individual risk factors. Revista do Instituto de Medicina Tropical de Sao Paulo 1995 May—June;37(3):225—33.

[70] Shriram AN, Murhekar MV, Ramaiah KD, Sehgal SC. Prevalence of diurnally subperiodic bancroftian filariasis among the Nicobarese in Andaman and Nicobar Islands, India: effect of age and gender. Tropical Medicine and International Health November 2002;7(11):949—54.

[71] Chesnais CB, Missamou F, Pion SD, Bopda J, Louya F, Majewski AC, et al. A case study of risk factors for lymphatic filariasis in the Republic of Congo. Parasites and Vectors July 1, 2014;7:300.

[72] Masriadi. The epidemiology and risk factor of lymphatic filariasis strains of *Wuchereria bancrofti* in Indonesia. Health Notions January 2018;2(1):40—4.

[73] Lammie PJ, Hitch WL, Walker Allen EM, Hightower W, Eberhard ML. Maternal filarial infection as risk factor for infection in children. Lancet April 27, 1991;337(8748):1005—6.

[74] Hightower AW, Lammie PJ, Eberhard ML. Maternal filarial infection — a persistent risk factor for microfilaremia in offspring? Parasitology Today November 1993;9(11):418—21.

[75] Malhotra I, Ouma JH, Wamachi A, Kioko J, Mungai P, Njovu M, et al. Influence of maternal filariasis on childhood infection and immunity to *Wuchereria bancrofti* in Kenya. Infection and Immunity September 2003;71(9):5231−7.

[76] Achary KG, Mandal NN, Mishra S, Mishra R, Sarangi SS, Satapathy AK, et al. *In utero* sensitization modulates IgG isotype, IFN-γ and IL-10 responses of neonates in bancroftian filariasis. Parasite Immunology October 2014;36(10):485−93.

[77] Bal MS, Mandal NN, DAS MK, Kar SK, Sarangi SS, Beuria MK. Transplacental transfer of filarial antigens from *Wuchereria bancrofti*-infected mothers to their offspring. Parasitology April 2010;137(4):669−73.

[78] Bal M, Sahu PK, Mandal N, Satapathy AK, Ranjit M, Kar SK. Maternal infection is a risk factor for early childhood infection in filariasis. PLoS Neglected Tropical Diseases July 30, 2015;9(7):e0003955.

[79] Braga C, Albuquerque MF, Schindler HC, Silva MR, Maciel A, Furtado A, et al. Risk factors for the occurrence of bancroftian filariasis infection in children living in endemic areas of northeast of Brazil. Journal of Tropical Pediatrics April 1998; 44(2):87−91.

[80] Alexander ND, Kazura JW, Bockarie MJ, Perry RT, Dimber ZB, Grenfell BT, et al. Parental infection confounded with local infection intensity as risk factors for childhood microfilaraemia in bancroftian filariasis. Transactions of the Royal Society of Tropical Medicine and Hygiene 1998 January−February;92(1):23−4.

[81] Weil GJ, Ramzy RM, El Setouhy M, Kandil AM, Ahmed ES, Faris R. A longitudinal study of Bancroftian filariasis in the Nile Delta of Egypt: baseline data and one-year follow-up. The American Journal of Tropical Medicine and Hygiene July 1999; 61(1):53−8.

[82] Gyapong JO, Adjei S, Sackey SO. Descriptive epidemiology of lymphatic filariasis in Ghana. Transactions of the Royal Society of Tropical Medicine and Hygiene 1996 January−February;90(1):26−30.

[83] Upadhyayula SM, Mutheneni SR, Kadiri MR, Kumaraswamy S, Nagalla B. A cohort study of lymphatic filariasis on socio economic conditions in Andhra Pradesh, India. PLoS One 2012;7(3):e33779.

[84] Sapada IE, Anwar C, Salni, Priadi DP. Environmental and socioeconomics factors associated with cases of cinical filariasis in banyuasin district of south sumatera, Indonesia. International Journal of Collaborative Research on Internal Medicine and Public Health 2015;7(6):132−40.

[85] Maryen Y, Kusnanto H, Indriani C. Risk factors of lymphatic filariasis in Manokwari, West Papua. TMJ 2017;4(1):60−4.

[86] Ikhwan Z, Herawati L, Suharti. Environmental, behavioral factors and filariasis incidence in bintan district, riau islands province. Kesmas: National Public Health Journal August 2016;11(1):39−45.

[87] Richards SL, Agada N, Balanay JAG, White AV. Permethrin treated clothing to protect outdoor workers: evaluation of different methods for mosquito exposure against populations with differing resistance status. Pathogens and Global Health February 2018; 112(1):13−21.

[88] Achee NL, Bangs MJ, Farlow R, Killeen GF, Lindsay S, Logan JG, et al. Spatial repellents: from discovery and development to evidence-based validation. Malaria Journal May 14, 2012;11:164.

[89] Scholte EJ, Knols BG, Samson RA, Takken W. Entomopathogenic fungi for mosquito control: a review. Journal of Insect Science 2004;4:19.

[90] Achee NL, Grieco JP, Vatandoost H, Seixas G, Pinto J, Ching-Ng L, et al. Alternative strategies for mosquito-borne arbovirus control. PLoS Neglected Tropical Diseases January 3, 2019;13(1):e0006822.

[91] Sunish IP, Kalimuthu M, Kumar VA, Munirathinam A, Nagaraj J, Tyagi BK, et al. Can community-based integrated vector control hasten the process of LF elimination? Parasitology Research June 2016;115(6):2353—62.

[92] Biritwum NK, Yikpotey P, Marfo BK, Odoom S, Mensah EO, Asiedu O, et al. Persistent 'hotspots' of lymphatic filariasis microfilaraemia despite 14 years of mass drug administration in Ghana. Transactions of the Royal Society of Tropical Medicine and Hygiene December 1, 2016;110(12):690—5.

[93] Molyneux DH, Taylor MJ. Current status and future prospects of the Global Lymphatic Filariasis Programme. Current Opinion in Infectious Diseases April 2001;14(2): 155—9.

[94] Gyapong JO, Kumaraswami V, Biswas G, Ottessen EA. Treatment strategies underpinning the global programme to eliminate lymphatic filariasis. Expert Opinion on Pharmacotherapy February 2005;6(2):179—200.

[95] World Health Organization. 2019. https://www.who.int/news-room/fact-sheets/detail/lymphatic-filariasis.

[96] Thomsen EK, Sanuku N, Baea M, Satofan S, Maki E, Lombore B, et al. Efficacy, safety, and pharmacokinetics of coadministered diethylcarbamazine, albendazole, and ivermectin for treatment of bancroftian filariasis. Clinical Infectious Diseases February 1, 2016;62(3):334—41.

[97] King CL, Suamani J, Sanuku N, Cheng YC, Satofan S, Mancuso B, et al. A trial of a triple-drug treatment for lymphatic filariasis. New England Journal of Medicine November 8, 2018;379(19):1801—10.

[98] Mukhopadhyay AK, Patnaik SK. Effect of mass drug administration programme on microfilaria carriers in East Godavari district of Andhra Pradesh. Journal of Vector Borne Diseases December 2007;44(4):277—80.

[99] Nandha B, Sadanandane C, Jambulingam P, Das P. Delivery strategy of mass annual single dose DEC administration to eliminate lymphatic filariasis in the urban areas of Pondicherry, South India: 5 years of experience. Filaria Journal August 24, 2007; 6:7.

[100] World Health Organization. Lymphatic filariasis transmission assessment surveys. 2011. p. 1—100. Geneva.

[101] Adhikari RK, Sherchand JB, Mishra SR, Ranabhat K, Devkota P, Mishra D, et al. Factors determining noncompliance to mass drug administration for lymphatic filariasis elimination in endemic districts of Nepal. Journal of Nepal Health Research Council 2014 May—August;12(27):124—9.

[102] Tangkawattana S, Sripa B. Integrative EcoHealth/One Health approach for sustainable liver fluke control: the Lawa model. Advances in Parasitology 2018;102:115—39.

[103] Mukhopadhyay AK, Patnaik SK, Satya Babu P, Rao KN. Knowledge on lymphatic filariasis and mass drug administration (MDA) programme in filaria endemic districts of Andhra Pradesh, India. Journal of Vector Borne Diseases March 2008;45(1):73—5.

[104] Haldar A, Dasgupta U, Ray RP, Jha SN, Haldar S, Bhattacharya SK. Critical appraisal of mass DEC compliance in a district of West Bengal. Journal of Communication Disorders 2013 Mar—June;45(1—2):65—72.

[105] Jones C, Tarimo DS, Malecela MN. Evidence of continued transmission of *Wuchereria bancrofti* and associated factors despite nine rounds of ivermectin and albendazole mass drug administration in Rufiji district, Tanzania. Tanzania Journal of Health Research April 2015;17(2):1—9.

[106] Jambulingam P, Subramanian S, de Vlas SJ, Vinubala C, Stolk WA. Mathematical modelling of lymphatic filariasis elimination programmes in India: required duration of mass drug administration and post-treatment level of infection indicators. Parasites and Vectors September 13, 2016;9:501.

[107] Burkot TR, Durrheim DN, Melrose WD, Speare R, Ichimori K. The argument for integrating vector control with multiple drug administration campaigns to ensure elimination of lymphatic filariasis. Filaria Journal August 16, 2006;5:10.

[108] Weerasooriya MV, Yahathugoda CT, Wickramasinghe D, Gunawardena KN, Dharmadasa RA, Vidanapathirana KK, et al. Social mobilisation, drug coverage and compliance and adverse reactions in a mass drug administration (MDA) programme for the elimination of lymphatic filariasis in Sri Lanka. Filaria Journal November 15, 2007;6:11.

[109] Supali T, Djuardi Y, Lomiga A, Nur Linda S, Iskandar E, Goss CW, et al. Comparison of the impact of annual and semiannual mass drug administration on lymphatic filariasis prevalence in Flores island, Indonesia. The American Journal of Tropical Medicine and Hygiene February 2019;100(2):336—43.

[110] Yates JA, Higashi GI. *Brugia malayi*: vaccination of jirds with [60]cobalt-attenuated infective stage larvae protects against homologous challenge. The American Journal of Tropical Medicine and Hygiene November 1985;34(6):1132—7.

[111] Paul R, Ilamaran M, Khatri V, Amdare N, Reddy MVR, Kaliraj P. Immunological evaluation of fusion protein of *Brugia malayi* abundant larval protein transcript-2 (BmALT-2) and Tuftsin in experimental mice model. Parasite Epidemiol Control February 7, 2019;4:e00092.

[112] Anand SB, Murugan V, Prabhu PR, Anandharaman V, Reddy MV, Kaliraj P. Comparison of immunogenicity, protective efficacy of single and cocktail DNA vaccine of *Brugia malayi* abundant larval transcript (ALT-2) and thioredoxin peroxidase (TPX) in mice. Acta Tropica August 2008;107(2):106—12.

[113] Khatri V, Chauhan N, Vishnoi K, von Gegerfelt A, Gittens C, Kalyanasundaram R. Prospects of developing a prophylactic vaccine against human lymphatic filariasis-evaluation of protection in non-human primates. International Journal for Parasitology August 2018;48(9—10):773—83.

[114] Coakley G, Maizels RM, Buck AH. Exosomes and Other Extracellular Vesicles: the new communicators in parasite infections. Trends in Parasitology October 2015; 31(10):477—89.

[115] Zamanian M, Fraser LM, Agbedanu PN, Harischandra H, Moorhead AR, Day TA, et al. Release of small RNA-containing exosome-like vesicles from the human filarial parasite *Brugia malayi*. PLoS Neglected Tropical Diseases September 24, 2015;9(9): e0004069.

[116] Harischandra H, Yuan W, Loghry HJ, Zamanian M, Kimber MJ. Profiling extracellular vesicle release by the filarial nematode *Brugia malayi* reveals sex-specific differences in cargo and a sensitivity to ivermectin. PLoS Neglected Tropical Diseases April 16, 2018;12(4):e0006438.

[117] Coakley G, McCaskill JL, Borger JG, Simbari F, Robertson E, Millar M, et al. Extracellular vesicles from a helminth parasite suppress macrophage activation and constitute an effective vaccine for protective immunity. Cell Reports May 23, 2017;19(8): 1545—57.

[118] Mekonnen GG, Pearson M, Loukas A, Sotillo J. Extracellular vesicles from parasitic helminths and their potential utility as vaccines. Expert Review of Vaccines March 2018;17(3):197−205.

[119] Debrah AY, Mand S, Marfo-Debrekyei Y, Batsa L, Pfarr K, Buttner M, et al. Macro-filaricidal effect of 4 weeks of treatment with doxycycline on *Wuchereria bancrofti*. Tropical Medicine and International Health December 2007;12(12):1433−41.

[120] Hoerauf A, Specht S, Büttner M, Pfarr K, Mand S, Fimmers R, et al. *Wolbachia* endo-bacteria depletion by doxycycline as antifilarial therapy has macrofilaricidal activity in onchocerciasis: a randomized placebo-controlled study. Medical Microbiology and Immunology September 2008;197(3):295−311.

[121] Sungpradit S, Chatsuwan T, Nuchprayoon S. Susceptibility of *Wolbachia*, an endo-symbiont of *Brugia malayi* microfilariae, to doxycycline determined by quantitative PCR assay. Southeast Asian Journal of Tropical Medicine and Public Health July 2012;43(4):841−50.

[122] Turner JD, Sharma R, Al Jayoussi G, Tyrer HE, Gamble J, Hayward L, et al. Alben-dazole and antibiotics synergize to deliver short-course anti-*Wolbachia* curative treat-ments in preclinical models of filariasis. Proceedings of the National Academy of Sciences of the United States of America November 7, 2017;114(45):E9712−21.

[123] Sharma R, Al Jayoussi G, Tyrer HE, Gamble J, Hayward L, Guimaraes AF, et al. Min-ocycline as a re-purposed anti-*Wolbachia* macrofilaricide: superiority compared with doxycycline regimens in a murine infection model of human lymphatic filariasis. Scientific Reports March 21, 2016;6:23458.

[124] Aljayyoussi G, Tyrer HE, Ford L, Sjoberg H, Pionnier N, Waterhouse D, et al. Short-course, high-dose rifampicin achieves *Wolbachia* depletion predictive of curative outcomes in preclinical models of lymphatic filariasis and onchocerciasis. Scientific Reports March 16, 2017;7(1):210.

[125] Johnston KL, Ford L, Taylor MJ. Overcoming the challenges of drug discovery for neglected tropical diseases: the A·WOL experience. Journal of Biomolecular Screening 2014 March;19(3):335−43.

[126] Johnston KL, Ford L, Umareddy I, Townson S, Specht S, Pfarr K, et al. Repurposing of approved drugs from the human pharmacopoeia to target *Wolbachia* endosymbionts of onchocerciasis and lymphatic filariasis. International Journal for Parasitology: Drugs and Drug Resistance September 16, 2014;4(3):278−86.

[127] Taylor MJ, Hoerauf A, Townson S, Slatko BE, Ward SA. Anti-*Wolbachia* drug discov-ery and development: safe macrofilaricides for onchocerciasis and lymphatic filariasis. Parasitology January 2014;141(1):119−27.

[128] Clare RH, Cook DA, Johnston KL, Ford L, Ward SA, Taylor MJ. Development and validation of a high-throughput anti-*Wolbachia* whole-cell screen: a route to macrofilaricidal drugs against onchocerciasis and lymphatic filariasis. Journal of Biomolecular Screening January 2015;20(1):64−9.

[129] Clare RH, Clark R, Bardelle C, Harper P, Collier M, Johnston KL, et al. Development of a high-throughput cytometric screen to identify anti-*Wolbachia* compounds: the po-wer of public-private partnership. SLAS Discovery 2019 June;24(5):537−47.

[130] Clare RH, Bardelle C, Harper P, Hong WD, Börjesson U, Johnston KL, et al. Industrial scale high-throughput screening delivers multiple fast acting macrofilaricides. Nature Communications January 2, 2019;10(1):11.

[131] O'Neill M, Mansour A, DiCosty U, Geary J, Dzimianski M, McCall SD, et al. An *in vitro/in vivo* model to analyze the effects of flubendazole exposure on adult female *Brugia malayi*. PLoS Neglected Tropical Diseases May 4, 2016;10(5):e0004698.

[132] O'Neill M, Ballesteros C, Tritten L, Burkman E, Zaky WI, Xia J, et al. Profiling the macrofilaricidal effects of flubendazole on adult female *Brugia malayi* using RNAseq. International Journal of Parasitology: Drugs and Drug Resistance December 2016;6(3):288—96.

[133] Sjoberg HT, Pionnier N, Aljayyoussi G, Metuge HM, Njouendou AJ, Chunda VC, et al. Short-course, oral flubendazole does not mediate significant efficacy against *Onchocerca* adult male worms or *Brugia* microfilariae in murine infection models. PLoS Neglected Tropical Diseases January 16, 2019;13(1):e0006356.

[134] Fischer C, Ibiricu Urriza I, Bulman CA, Lim KC, Gut J, Lachau-Durand S, et al. Efficacy of subcutaneous doses and a new oral amorphous solid dispersion formulation of flubendazole on male jirds (*Meriones unguiculatus*) infected with the filarial nematode *Brugia pahangi*. PLoS Neglected Tropical Diseases January 16, 2019;13(1):e0006787.

[135] Vialpando M, Smulders S, Bone S, Jager C, Vodak D, Van Speybroeck M, et al. Evaluation of three amorphous drug delivery technologies to improve the oral absorption of flubendazole. Journal of Pharmaceutical Sciences September 2016;105(9):2782—93.

[136] Tuerk C, Gold L. Systematic evolution of ligands by exponential enrichment: RNA ligands to bacteriophage T4 DNA polymerase. Science August 3, 1990;249(4968):505—10.

[137] Long Y, Qin Z, Duan M, Li S, Wu X, Lin W, et al. Screening and identification of DNA aptamers toward *Schistosoma japonicum* eggs via SELEX. Scientific Reports April 28, 2016;6:24986.

[138] Ulrich H, Magdesian MH, Alves MJ, Colli W. *In vitro* selection of RNA aptamers that bind to cell adhesion receptors of *Trypanosoma cruzi* and inhibit cell invasion. Journal of Biological Chemistry June 7, 2002;277(23):20756—62.

[139] Homann M, Göringer HU. Uptake and intracellular transport of RNA aptamers in African trypanosomes suggest therapeutic "piggy-back" approach. Bioorganic and Medicinal Chemistry October 2001;9(10):2571—80.

[140] Moreno M, González VM. Advances on aptamers targeting Plasmodium and trypanosomatids. Current Medicinal Chemistry 2011;18(32):5003—10.

[141] Song KM, Lee S, Ban C. Aptamers and their biological applications. Sensors 2012; 12(1):612—31.

[142] Ospina-Villa JD, Zamorano-Carrillo A, Castañón-Sánchez CA, Ramírez-Moreno E, Marchat LA. Aptamers as a promising approach for the control of parasitic diseases. Brazilian Journal of Infectious Diseases 2016 November—December;20(6):610—8.

[143] Andrade RM, Reed SL. New drug target in protozoan parasites: the role of thioredoxin reductase. Frontiers in Microbiology September 30, 2015;6:975.

[144] Onodera T, Momose I, Kawada M. Potential anticancer activity of auranofin. Chemical and Pharmaceutical Bulletin 2019;67(3):186—91.

[145] Bulman CA, Bidlow CM, Lustigman S, Cho-Ngwa F, Williams D, Rascón Jr AA, et al. Repurposing auranofin as a lead candidate for treatment of lymphatic filariasis and onchocerciasis. PLoS Neglected Tropical Diseases February 20, 2015;9(2):e0003534.

[146] Rao UR, Salinas G, Mehta K, Klei TR. Identification and localization of glutathione S-transferase as a potential target enzyme in *Brugia* species. Parasitology Research 2000 November;86(11):908—15.

[147] Saeed M, Imran M, Baig MH, Kausar MA, Shahid S, Ahmad I. Virtual screening of natural anti-filarial compounds against glutathione-S-transferase of *Brugia malayi* and *Wuchereria bancrofti*. Cellular and Molecular Biology October 30, 2018; 64(13):69−73.

[148] Choi TW, Cho JH, Ahnn J, Song HO. Novel Findings of anti-filarial drug target and structure-based virtual screening for drug discovery. International Journal of Molecular Sciences November 13, 2018;19(11):3579.

[149] Amala M, Rajamanikandan S, Prabhu D, Surekha K, Jeyakanthan J. Identification of anti-filarial leads against aspartate semialdehyde dehydrogenase of *Wolbachia* endosymbiont of *Brugia malayi*: combined molecular docking and molecular dynamics approaches. Journal of Biomolecular Structure and Dynamics February 2019;37(2): 394−410.

[150] Ballesteros C, Geary JF, Mackenzie CD, Geary TG. Characterization of divalent metal transporter 1 (DMT1) in *Brugia malayi* suggests an intestinal-associated pathway for iron absorption. International Journal of Parasitology: Drugs and Drug Resistance August 2018;8(2):341−9.

[151] Wu B, Novelli J, Foster J, Vaisvila R, Conway L, Ingram J, et al. The heme biosynthetic pathway of the obligate *Wolbachia* endosymbiont of *Brugia malayi* as a potential anti-filarial drug target. PLoS Neglected Tropical Diseases July 14, 2009;3(7): e475.

[152] Li Z, Carlow CK. Characterization of transcription factors that regulate the type IV secretion system and riboflavin biosynthesis in *Wolbachia* of *Brugia malayi*. PLoS One 2012;7(12):e51597.

[153] Carpinone EM, Li Z, Mills MK, Foltz C, Brannon ER, Carlow CKS, et al. Identification of putative effectors of the type IV secretion system from the *Wolbachia* endosymbiont of *Brugia malayi*. PLoS One September 27, 2018;13(9):e0204736.

[154] Pallen MJ. Diagnostic metagenomics: potential applications to bacterial, viral and parasitic infections. Parasitology December 2014;141(14):1856−62.

[155] Bender JM, Dien Bard J. Metagenomics in pediatrics: using a shotgun approach to diagnose infections. Current Opinion in Pediatrics February 2018;30(1):125−30.

[156] Gao D, Yu Q, Wang G, Wang G, Xiong F. Diagnosis of a malayan filariasis case using a shotgun diagnostic metagenomics assay. Parasites and Vectors February 16, 2016;9:86.

[157] Parashar A. Aptamer and its role in diagnostics. International Journal of Bioassays 2016;5(2):4799−809.

[158] Sundararaj N, Kalagatur NK, Mudili V, Krishna K, Antonysamy M. Isolation and identification of enterotoxigenic *Staphylococcus aureus* isolates from Indian food samples: evaluation of in-house developed aptamer linked sandwich ELISA (ALISA) method. Journal of Food Science and Technology February 2019;56(2): 1016−26.

[159] Ramos E, Piñeiro D, Soto M, Abanades DR, Martín ME, Salinas M, et al. A DNA aptamer population specifically detects *Leishmania infantum* H2A antigen. Laboratory Investigation May 2007;87(5):409−16.

[160] Guerra-Pérez N, Ramos E, García-Hernández M, Pinto C, Soto M, Martín ME, et al. Molecular and functional characterization of ssDNA aptamers that specifically bind *Leishmania infantum* PABP. PLoS One October 12, 2015;10(10):e0140048.

[161] Li F, Yu Z, Han X, Lai RY. Electrochemical aptamer-based sensors for food and water analysis: a review. Analytica Chimica Acta March 21, 2019;1051:1−23.

[162] Chen A, Yang S. Replacing antibodies with aptamers in lateral flow immunoassay. Biosensors and Bioelectronics September 15, 2015;71:230—42.

[163] Kaiser L, Weisser J, Kohl M, Deigner HP. Small molecule detection with aptamer based lateral flow assays: applying aptamer-C-reactive protein cross-recognition for ampicillin detection. Scientific Reports April 4, 2018;8(1):5628.

[164] Frohnmeyer E, Tuschel N, Sitz T, Hermann C, Dahl GT, Schulz F, et al. Aptamer lateral flow assays for rapid and sensitive detection of cholera toxin. Analyst February 25, 2019;144(5):1840—9.

[165] Stutzer C, Richards SA, Ferreira M, Baron S, Maritz-Olivier C. Metazoan parasite vaccines: present status and future prospects. Frontiers in Cellular and Infection Microbiology March 13, 2018;8:67.

Malaria

5

Nikunj Tandel[1], Rajeev K. Tyagi[2]

[1]*Institute of Science, Nirma University, Ahmedabad, Gujarat, India;* [2]*Ramalingaswami Fellow and Faculty, Division of Cell Biology and Immunology, Biomedical Parasitology and Nano-immunology Lab, CSIR-Institute of Microbial Technology (IMTECH), Chandigarh, India*

5.1 Introduction

The major challenge for any developing (as well for developed) country around the globe is to control or eliminate the burden of any "infectious diseases." As these infectious diseases are the leading causes of death and disability and halted the human progress through other numerous synergistic challenges [1]. The emergences of new, unrecognized, and epidemic impact of old infectious diseases continuously present a formidable challenge. In the past three and half decades, more than 30 new infectious diseases have emerged and added an additional burden to the existing epidemic outbreaks of AIDS, malaria, Pandemic Influenza, Ebola, Dengue, and many more [1,2]. Among all the listed and other infectious diseases, malaria, vector-borne parasitic diseases still remain the major global health challenge found in almost 90 counties across the world [3,4]. It is a life-threatening disease caused by the parasites of *Plasmodium* family and transmitted to the people through the bite of the female *anopheles* mosquitoes that carry the infectious sporozoites [5]. It is an ancient disease documented and found in several historical documents. It has an *Italian* origin (mal'aria) means the "spoiled air" [6]. An estimated 219 million cases were reported for malaria around the world in 2017 as compared to 239 and 217 million cases in the 2010 and 2016, respectively. Despite a decrease of 20 million malaria cases, 11th world malaria report suggested that the progress of reduction of malaria burden from the data of 2015 to 2017 is stalled [4]. The total death toll due to malaria reached 4,35,000 in the last year, and children under 5 years of the age is the most vulnerable group that accounted for 61% of total death [4]. At a glance, majority of the cases were registered from Africa region (92%) followed by South-East Asia (5%) and Eastern Mediterranean (2%) and out of total death, 80% of the death were reported from 17 countries of Africa and India [4]. Out of more than 120 species of *Plasmodium* which infects the reptiles, birds, and mammals, currently, six species are known to infect humans, lead by *Plasmodium falciparum*, most lethal malaria parasite that causes cerebral malaria at the advanced stage followed by *Plasmodium vivax* [3,4]. As per the *Global Technical Strategy (GTS) for Malaria 2016–30*, the target of curtailing down the global malaria incidence and mortality rate is set to 90% by 2030 [7]. The current status of malaria cases and their distribution in 2017 are illustrated in Fig. 5.1.

This systemic disease, its pathophysiology, and molecular mechanisms, time to time, have been revealed by researchers, which further adds on to improve our knowledge of malaria biology. The successful culture of *Plasmodium* has been achieved *in vitro* by Tiger and Jansen in 1976, which served as a blueprint for the malaria biologist to understand the host–pathogen interaction in more details [5]. Although it remains an unsolved puzzle, as vector control and chemoprophylaxis/chemotherapy, the two major weapons against malaria are unsuccessful because of drug resistance by the parasite [8] and insecticide resistance by the mosquito vector [9]. The reports of failure of frontline antimalarial drugs and their combination therapy in the last decade against the various *plasmodia* strains is perturbing as it frightens to make the malaria practically untreatable globally and eliminate the disease [10,11].

The vaccine development is another aspect that has gained significant attraction, as devising an antimalaria vaccine can not only help in overcoming the problem of multiple drug resistance but also in capping the morbidity associated with the disease [12]. The vaccine might prove panacea in fighting malaria infection; however, an effective malaria vaccine is still required owing to several bottlenecks [13].

5.1.1 Malaria life cycle: understanding the *Plasmodium* biology and disease

The complete malaria life cycle in both the host; humans (intermediate host) and mosquito (definitive host) has been reviewed in detail by others [3,6,14–18], and therefore we have summarized the glimpse. Malaria infection in humans begins when the infected sporozoites from the salivary glands of the female *Anopheles* mosquito gain entry to the bloodstream through which they travel to the liver and undergoes tight regulation before invading the hepatocytes [3]. After having invaded the hepatocytes, these sporozoites replicate leading to formation of schizonts; this stage of asexual reproduction inside the liver cells usually lasts for about a week. Each schizont gives rise to several merozoites that are released into the bloodstream, marking the end of the exoerythrocytic phase [17]. Infection by *P. vivax* and *P. ovale* is not characterized with this reproduction step (the pathogens may exist as hypnozoites in the hepatocytes for a long duration until relapse occurs, releasing merozoites into the bloodstream) [14,19]. The merozoites in the bloodstream invade erythrocytes followed by the formation of a ring that later evolves into a trophozoite and finally develops into erythrocytic schizont [14]. Each mature erythrocytic schizont gives rise to a generation of new merozoites that eventually rupture erythrocytes and are released in the bloodstream to invade other RBCs. This is the stage when the clinical manifestations of the disease begins to appear. Unlike the liver stage, the erythrocytic stage of the lifecycle continues multiple times [16,20].

Another development that takes place in the RBCs is the differentiation of the parasite into male and female gametocytes, the nonpathogenic forms. Now, during

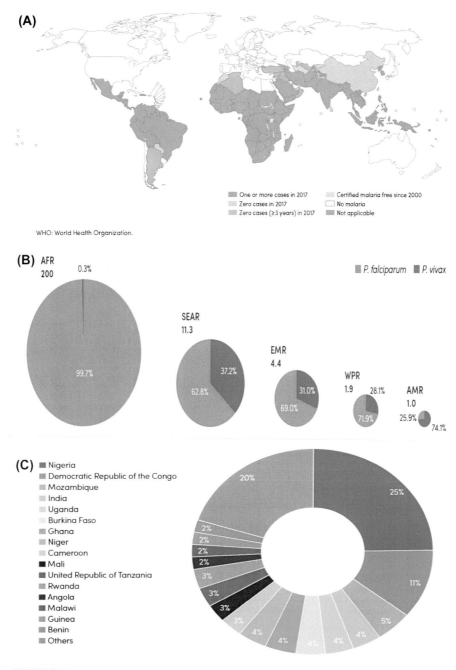

FIGURE 5.1

(A) Current scenario of malaria indigenous cases in 2000 and their status in 2017, (B) estimated malaria cases by WHO regions, and (C) country-wise distribution of total malaria cases.

Adapted with permission from WHO. World malaria report 2018: World Health Organization; 2018.

this stage, if a female *Anopheles* mosquito takes a blood meal, the gametocytes are taken up and mature into macrogametes (female) and microgametes (male) [15,20]. In the gut of the mosquito, the microgametes undergoes three divisions, producing eight nuclei in the process. Each nucleus fertilizes with a macrogamete resulting in the formation of an ookinete. The ookinete penetrates the midgut wall and gets encapsulated, forming an oocyst [15,16]. Inside the oocyst, the ookinete nucleus divides, producing thousands of sporozoites. Toward, the end of sporogony that lasts 8–15 days, the ookinete ruptures and the sporozoites travel to the salivary gland of the mosquito, ready to be injected into a human host during the next blood meal of the mosquito [13,18,21]. The detailed basic malaria life cycle of *Plasmodium* species has been depicted in Fig. 5.2.

5.2 Detection of infection

The success of any health system for the management of the malaria cases mainly relies on the extent of malaria-suspected patients seeking treatment followed by care, appropriate and prompt diagnosis, and, if tested positive for malaria, treated properly [4]. The burden of malaria has propelled the interest in developing the cost-effective diagnostic strategies in the developing world as well as in the developed world where it lacks the expertize in identification [22]. The detection of malaria infection is basically examination of the blood sample for the presence of parasite antigen or its alternative metabolite products [23]. The sign and symptoms of malaria patients are moreover similar to the feverish illness which includes the headache, sweat, shivering, muscle pain, fatigue, nausea, and occasionally vomiting [4]. To identify whether the patients truely suffer due to malaria or not is mainly clinic-based followed by WHO recommended tests such as rapid diagnostic tests (RDT) or the gold standard method of microscopic examination [23]. However, different parasitic species, geographical locations, atypical vector, and their transmission, level of parasitemia, migrants, recurrence and sequestration into deep-seated tissues, drug resistance, and treatment based on assumption of the patients, have hampered the malaria eradication efforts [22,23]. To scrutinize the parasite recognition, blood is required, which is a taboo subject; moreover, in the case of children under the age of 5 years, it is more difficult to diagnose. Alongside, the basic identification methods, RDT and microscopic detection has the limitation of false-positive results and less sensitive for low level of parasitemia, respectively [23]. Nevertheless, the advanced techniques such as polymeric chain reaction (PCR), microarray, flow cytometry assay, and others are highly sensitive and currently in use [24–26], still the cost, time, and the requirement of the expertise make them unfavorable for the routine usages in remote endemic areas where the delay in decision may be fatal for the patients [27]. To overcome the inadequacy of existing identification methods, recently saliva [28–31], urine [32] or both together (saliva and urine) [33,34], and body secretion and surface mucosa [35] are in use with the combination of other existing methods. Currently, the latest

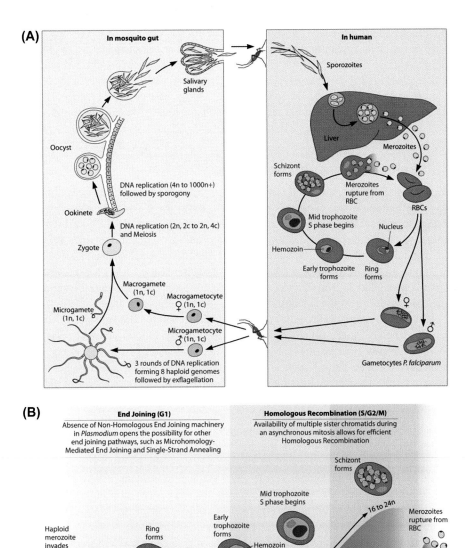

FIGURE 5.2

(A) The complete malaria life cycle of *Plasmodium* species and (B) the intraerythrocytic asexual *Plasmodium* cycle in the human host.

Adapted with permission from Lee AH, Symington LS, Fidock DA. DNA repair mechanisms and their biological roles in the malaria parasite Plasmodium falciparum. *Microbiol Mol Biol Rev 2014;78(3):469–486.*

technology mainly focuses on developing a point-of-care (POC)-based diagnostic test for the field application to improve all the test parameters; it can bridge the gap between a clinician and diagnostic laboratory in the resource-limited malaria-endemic areas [23]. Simple RDT and microscopy to sophisticated PCR technique and others are currently in use and all of them have their own advantages and disadvantages. The different identification methods have been listed in Table 5.1.

5.3 Epidemiology and risk factors

As mentioned earlier, although the number of malaria cases are reduced, the progress of elimination and control is halted throughout the world. Malaria is majorly found in tropical and subtropical regions and spread through the bite of female *Anopheles* mosquitoes [36]. The global distribution of malaria mainly relies on the mosquito vectors and out of 400 different *Anopheles* species around 31—40 vectors are called as "physiologically competent vectors" (Fig. 5.3) [4,37], which are capable enough to develop the infectious parasites despite their occurrence in experimental conditions in the laboratory [38].

This vector mainly bites between dusk and dawn and the transmission of disease mainly depends on the climate and environmental conditions together with local vector (mosquito) behavior [4]. Moreover, the incidence of malaria has also been influenced by other factors such as vector control, human host, and parasites which are strict anthroponoses without any animal reservoirs, and therefore it is indistinguishably linked to the poor and socioeconomic unstable countries, areas affected with natural calamities and war with inadequate resources [36]. Transmissions from mother to child or through the blood transfusion are few of the less common routes, however at higher risk in the area with restricted resources [36]. Out of all four major *Plasmodium* parasites, *P. falciparum* accounts for 99.7% malaria incidence in WHO African regions and *P. vivax* is the predominant in WHO American regions representing 74.1% malaria cases [4]. The other remaining species *P. malariae*, *P. ovale*, and newly emerged *P. knowlesi* from the Malaysia also accounts for malaria; however, the rate of incidence is low [36].

Several risk factors are associated with the malaria burden, and the main categories are (1) factors influencing malaria and (2) factors influencing the vector. The key factor that mainly affects the malaria prevalence is illustrated in Fig. 5.4.

5.3.1 Factor influencing malaria

The immunity at the individual level and the health status of the respective population plays an imperative role in malaria morbidity and mortality. To restrain the malaria infection fundamentally relies on the acquired immunity of individuals which is exposure and age-dependent [4]. In low transmission area, the prevalence of malaria is age-independent and moreover similar in all age groups; however, in moderate level transmission area, immunity will enhance with the increase of transmission

	Clinical diagnosis	PBS	QBC	RDTs	Serological tests	PCR	LAMP	Microarrays	FCM	ACC	MS	Biosensors
Principle	Observation of sign and symptoms	Observation of stained thick and thin blood smear under microscope	Staining of blood with acridine orange and observation under fluorescent microscope	Detection of parasite antigens/ enzymes/ metabolite products	Antibody detection	Identification through amplification of malaria DNA	Amplification of malaria DNA followed by turbidity measurement	Quantification and detection of DNA by hybridization	Flow cytometric identification of hemozoin	Malarial Pigment identification in activated monocytes	Detection of heme through LDMS	Analytical devices which can analyze the blood or urine without any additional processing or usage of reagent
Sensitivity and specificity	Depends on ability of clinic and expertise of clinician	Requirement of expert personnel alongside laboratory requirements	Higher then PBS More sensitive for falciparum species then non-falciparum species Less specificity due to staining of leukocyte DNA	Moderate level if more than 100 parasites/µl & less sensitive to identify the <10 parasites	Higher up-to certain extends however unable to compare with clinical symptoms of patients	Highest sensitivity but required less specific to identify mix infection	Highest	Relatively higher	High specificity and sensitivity varies	Sensitivity and specificity varies Further studies can enhance the validity of instrument and software	Protocol under development	Under progress
Time duration (min)	N/A	30-60	<15	5-15 sometimes 15-30	30-60	60-360 (depends upon the method)	<60	<60	Machine based and maximum 1/sample	Machine based and maximum 1/sample	Machine based and maximum 1/sample	Based on the type of sensor and marker used for the identification
Detection limit (parasites/ µl)	N/A	Expert ~5-10 In routine > 50	>5	50-100	Undetermined	≥1	>5	Undetermined	Irrelevant correlation with parasitemia	5-20	10^4 µl of blood	1-2.5 parasites/ µl
Expertise	Higher	Higher specially in non-endemic area	Moderate,	Low	Moderate	High	High	High	High	High	High	High
Instrumental Cost	N/A	Low	Moderate	Moderate	Moderate	Expensive compare to RDT and microscopy otherwise less costly compare to ACC, MS, FC, etc	Moderate	Expensive	Expensive	Expensive	Expensive	Based on usage of chip they may be cheaper/expensive
Other Consideration	Identification of mixed infection and handling of severe malaria is still problematic	Gold standard method for human parasites except for *P. knowlesi*, identification of mixed infection and difficulties during low parasitemia	Requirement of electricity, unable to identify the parasites as well capillaries cannot be stored for further references	Currently 1st line of diagnostic method Unable to differentiate between *P. vivax*, *P. ovale* & *P. malariae*	The results can be influenced by trained personnel, useful in identification of the infection however not for the treatment purpose	2nd line of diagnostics in most of the laboratories, helpful in species identification, parasite quantifications, helpful in identification the drug-resistance markers	Limited in quantification of the parasites however reliability and its usages required further validation	In the initial stage of development for diagnosis of the malaria	Useful in diagnosis of unsuspected clinical malaria; Further validation is required for full usage	Clinical trials are indeed to validate its feasibility and utility	In the initial stage of development for diagnosis of the malaria	Certain biosensors can suitable for diagnosis of malaria at low parasitemia while other can make the difference between *P.falciparum* and *P.vivax* Sensors can also be stored at room temperature up-to 2 months

PBS, peripheral blood smear; QBC, quantitative buffy coat; RDTs, rapid diagnostic tests; PCR, polymerase chain reaction; LAMP, loop-mediated isothermal amplification; FCM, flow cytometry; ACC, automated blood cell counter; MS, mass spectrometry; LDMS, laser desorption mass spectrometry

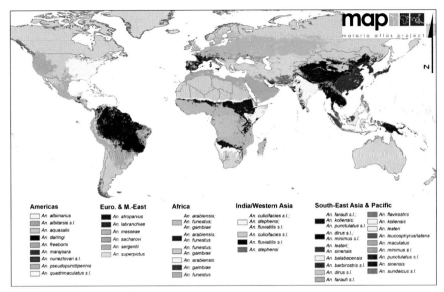

FIGURE 5.3

A global map of dominant malaria vector.

Adapted with permission from Sinka ME, Bangs MJ, Manguin S, et al. A global map of dominant malaria vectors.

Parasites and Vectors. 2012;5(1):69.

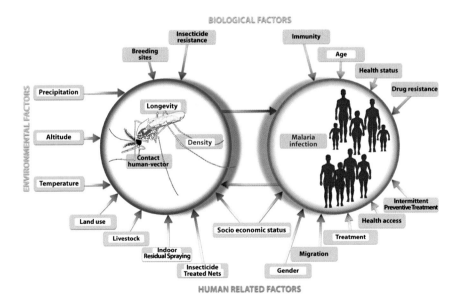

FIGURE 5.4

The list of risk factors that affect malaria prevalence.

Adapted with permission from Protopopoff N, Van Bortel W, Speybroeck N, et al. Ranking malaria risk factors to guide malaria control efforts in African highlands. PLoS One. 2009;4(11):e8022.

up-to certain age (2—3 years) and later on it will be independent of transmission level [40]. Pregnant women are more prone to malaria infections as compared to non-pregnant women in endemic areas. Similarly, malnutrition is also one of the crucial factors due to which, it mainly affects the children under 5 years of age [39]. The bleak situation of malaria reoccurrence in some of the areas is directly associated with the drug-resistance phenomena alongside the other human allied factors of poor health system access, quality, migrants and non-immune travelers and socio-economic growth of the place [4,39].

5.3.2 Factor influencing the vector

The effective transmission of malaria from one to another individual is determined by the mosquito vector's characteristics of abundance, longevity, capacity and direct contact with humans. On the other hand geographical altitude, a temperature that influences the longevity and feeding frequency of a mosquito, climate condition especially rainfall during which humidity will increase and above 60% of it favors the longevity of adult mosquitoes. However, heavy rainfall washes away all the larva of the vector that is also an important contributor to transmission [4,41]. The forest areas are the hot-beds for malaria transmission, as the topographic parameters such as humidity, temperature, rainfall, and vegetation provide the utmost favorable conditions for the vectors to survive [42].

5.4 Approaches to control and elimination through mass drug administration

Over the past decade, malaria elimination and eradication have augmented owing to various new diagnostic tools for the detection and treatment, vector control through insecticide-treated nets (ITNs) and indoor residual spraying (IRS), an improved surveillance mechanism, novel approaches of vaccines to prevent malaria and joint adventure at the global level by WHO and other firms with the target of "world free malaria" and curtail down the mortality rate up to 90% by 2030 [4,43]. There are more than 17 prequalified vector control products [44], around 25 malaria vaccines in the pipeline under the global malaria vaccine initiative [45], and almost 50 medicines are in their preclinical/clinical trials [43]. Despite the various approaches for the vector control, the situation of insecticide resistance has been accounted against almost all the anopheline vector globally [46] and approximately 68 countries have reported the insecticide resistance against 1 out of 5 commonly-used nets. Out of them, roughly 57 countries have detailed resistance against two or more insecticides (mainly against pyrethroids—the only insecticide used in ITNs) [46,47].

Similarly, over the last two centuries, the malarial parasite has gained the resistance starting from quinine, the first antimalarial drug to the latest frontline medicine of artemisinin combination therapy (ACT) in 2008, within the short duration of its usage since 2004 [48]. The unavailability of other alternatives to the ACT and recent report of artemisinin resistance in the Greater Mekong Subregion (GMS) has created an alarming situation to prevent the spread of resistance to the African region [48]. To surmount the multidrug resistance condition, scientists have again started focusing on mass drug administration (MDA), an earlier component of several malaria elimination programs of the mid-20th century; however, the issues related to its safety, efficacy, sustainability, feasibility, alarming condition of enhanced drug resistance and less goal orientation program have raised the questions for its continuity in several regions [49]. MDA is a time-limited administration of antimalarial drugs to the specific group of people irrespective of their status of illness at the similar time point, especially in the area where malaria is approaching elimination (Fig. 5.5) [50] and termed as targeted malaria elimination (TME) [51]. The target of asymptomatic infection which is generally unable to detect by the routinely used methods makes them the most successful method alongside ACT treatment. The selection of MDA is based on less evidence and from the area of less endemicity in WHO African settings, although the different model-based studies have provided certain guidance [50].

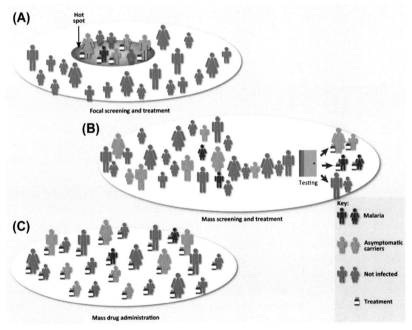

FIGURE 5.5

The schematic approach of MDA with other treatment approaches of mass screening, treatment (MSaT) and focal MDA or screening and treatment (FMAD/FSAT).

Aadapted with permission from Grueninger H, Hamed K. Transitioning from malaria control to elimination: the vital role of ACTs. Trends in Parasitology. 2013;29(2):60–64.

The potential role of MDA has been summarized by Newby *et al.*, which includes the coverage of almost 80%−90% people of the targeted population, direct drug administration, and easy management with strong engagement in the respective community alongside the control on the concomitant vector by IRS and ITNs [53]. As it mainly dampened the malaria transmission, there are several additional concerns on priorities. The selection criteria for pregnant women, children under 5 years of age, and emergencies of the outbreak of other infectious diseases are the factors to keep in mind before any MDA treatment [49,53]. Brady *et al.* and colleagues have studied several models of malaria transmission such as EMOD DTK, Imperial, MORU, and Open-Malaria and concluded that despite the individual malaria MDA program, the magnitude is different; the characteristics of the respective program and the location where the MDA program is going to be implemented play a crucial role in its success [54]. The successful completion of people enrollment throughout the one round of MDA per year and the total duration of the program has the major influence on the effectiveness of the MDA program [54]. In the recent trial conducted by Morris and colleagues in Zanzibar, dihydroartemisinin-piperaquine (DHAP) was used in two rounds of MDA together with single lower dose of primaquine (PQ) [55]. The results of the trials are sobering as it has no impact on malaria incidence and transient effect on the prevalence of asymptomatic infections confirmed by the PCR method. There are several loopholes in the optimization and implementation of the study although it has shown us the way to tackle the on-going transmission in near-to-elimination area and requirement of more in-depth investigations that pave the light on malaria epidemiology [50,55]. Kaehler *et al.* have suggested several factors (Fig. 5.6) which affect the implementation and acceptance of the MDA program concluded from an interview with several policymakers and principal investigators, actively involved in the malaria elimination program in GMS [48].

The recent cluster-randomized trial conducted by Seidlein *et al.* in Southeast Asia on *P. falciparum* with 3 monthly rounds of DHAP MDA over the 1 year period in the areas affected with artemisinin resistance had shown the reduction in the incidence and prevalence of falciparum malaria; however, the small sample and heterogeneity across the villages have restricted impact on the intervention [56]. All the current results of the MDA program suggests that it can hasten the falciparum malaria elimination if it is well-managed and well-organized with full resources [56]. Elimination and control of malaria transmission are mainly lead by accurate and proper medicine; however, the recent development of technology has gained the attention of malaria biologists for novel approaches of its diagnosis, treatment, and prevention.

5.5 Next-generation vaccines, drugs, and diagnostics

Despite the malaria parasite identification by Ronald Ross in 1897, it still remains as the major burden across the globe and almost US$ 3.1 billion has been spent behind the malaria elimination and control globally in 2017 [4]. The recent alarming

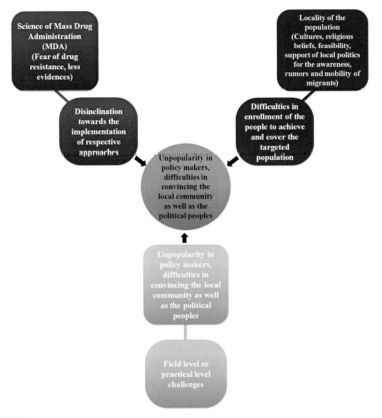

FIGURE 5.6

The list of factors affecting the implementation and acceptability of the MDA program.

Adapted and modified from Kaehler N, Adhikari B, Cheah PY, et al. The promise, problems and pitfalls of mass drug administration for malaria elimination: a qualitative study with scientists and policymakers. International Health. 2019;11(3):166–176.

condition of drug and insecticide resistance has created pressure on researchers for novel interventions and strategies for fast and accurate identification, especially for remote areas which are deprived of limited-resources and skilled personnel [57]. The key challenges of novel diagnostics are to identify the subclinical infection (asymptomatic malaria) and low level of parasitemia in endemic areas which enable and enhance the specificity of already in-use techniques for the appropriate surveillance system to eliminate malarial parasites [23,57]. An initiative of Malaria Eradication Research Agenda (maIERA) 2011 started with the aim to bridge the gap between the academician and researchers (more than 250 scientists from 36 or more countries) for the development of new diagnostics, drugs, vaccines, and vector control programs (Fig. 5.7) followed by another program in 2012 named Malaria Eradication Scientific Alliance funded by Bill and Melinda Gates Foundation (OPP1034591) [57].

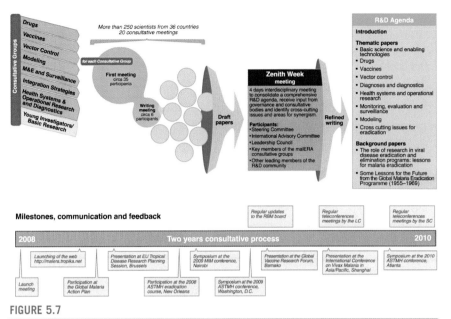

FIGURE 5.7

The agenda of research and development for malaria elimination.

Adapted with permission from Alonso PL, Brown G, Arevalo-Herrera M, et al. A research agenda to underpin malaria eradication. PLoS Medicine. 2011;8(1):e1000406.

Although the implementation of above mentioned and other projects, malaria elimination has not reached its target. This necessitates the requirement of novel tools and technologies along with the understanding of the mechanism in detail, for surveillance and spatial decision support system (SDSS) for more prominent future activities [43]. All the major new products are under development through Product Development Partnerships (PDPs) with various ventures such as Medicines for Malaria Venture (MMV), Malaria Vaccine Imitative (MVI), Program for Appropriate Technology in Health (PATH), Novartis Institute for Tropical Diseases (NITD), European Vaccine Initiative (EVI), and others [43]. Drugs or vaccines which can directly target the parasites at multiple levels than one single stage of its lifecycle, novel tools which can hinder the vector and their transmission and various diagnostic techniques and surveillance fall under the scope of more than 100 products that are in the development phase and some of them will be available soon in near future [43].

Drugs are developed under the program of Single Encounter Radical Cure and Prophylaxis (SERCaP) with the target product profile (TPP), which can kill all the stages of parasites including the resting hypnozoites of *P. vivax* or *P. ovale* in the liver, suitable for the MDA and prophylaxis for almost a month after treatment to study the parasite development in mosquito vector (Fig. 5.8) [43,57].

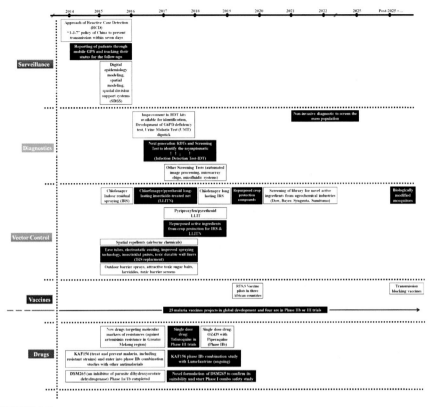

FIGURE 5.8

Timeline of currently in use and upcoming novel malaria diagnostics/identification methods.

Adapted and modified from Hemingway J, Shretta R, Wells TN, et al. Tools and strategies for malaria control and elimination: what do we need to achieve a grand convergence in malaria?. PLoS Biology. 2016;14(3): e1002380.

Vaccines which are currently in clinical trials mainly focus on reducing the morbidity and mortality of *P. falciparum* in children less than 5 years of age in the African region. Currently, vaccines which can hinder and interrupt the malaria transmission are also indeed therefore; broad concept of "vaccine that interrupts malaria transmission (VMIT)" is given by malERA to replace the term "transmission blocking vaccines (TBV)," which is used for the vaccine that targets only the sexual and mosquito stages of parasites (Fig. 5.8) [57]. VMIT focuses on the development of antivector vaccines that target the essential molecules required for the parasite development, vaccines for pre-erythrocytic and erythrocytic stages, and also the parasite antigens of both mosquito and sexual stages of the infection [57]. Recently, the first malaria vaccine RTS,S/AS01 (RTS,S) is launched in Africa in the pilot program, one of the breakthroughs in the vaccine development in the last 70 years [59].

Vector control is one of the major components of malaria elimination in highly endemic areas. Recently, to overcome the insecticide resistance issues, several other approaches such as the usage of noninsecticidal methods includes the genetic approach with bacterial endosymbionts [43], long-lasting insecticidal nets (LLINs), and long-lasting insecticide-treated hammocks for the remote area [57] are in use/under development. The management of the source of the larva and environmental factors, vector behavior and their outdoor targeting, electrostatic coating to enhance the bioavailability of insecticides, improved spraying technologies, coincide development with seasonal transmission and other ways can improve the vector control [57,58].

Diagnostic, if improved, can assist in measuring the changes in infection and strengthen the usage of medicine and treatment choices. All the existing RDTs for various parasites have their pros and cons [23]. Glucose-6-phosphate dehydrogenase (G6PD) test, next-generation RDTs and other screening tests (automated image processing, microarray chips, microfluidic systems, LAMP method), non blood-based assays (urine dipstick) [23], and detection of multiple biomarkers and others (biosensors) can enhance the specificity and sensitivity for low parasitemia [23,43].

Surveillance and Responses also play a vital role where malaria has reached to the elimination stage and used as a role model for other studies for case reporting, management, and compiling the recorded data in a representable manner that shortens the process of decision between various policymakers, healthcare professionals, and researchers [43].

5.6 Novel pathways of drug targets for malaria intervention

Over the last 6 decades, we are trying to find out the radical cure for malaria and achieved partial success by following the guidelines for malaria treatment approved by WHO [49]. At present, all the available antimalarial drugs are classified into three categories on the basis of their chemical structure and mode of action [60]. The metabolite differences of the malarial parasites with respect to host such as heme toxification, oxidative stress, nucleic acid metabolism, and fatty acid biosynthesis are some of the major pathways mainly targeted for the antimalarial drugs [61]. Although the severity and death toll has been halted due to malaria, the recent surge of drug resistance against the first line of defense is the major obstacle as almost all the antimalarial drugs are in practice since decades and none of them as been designed in a fully rational manner for known drug targets [62]. In addition to this, all of them have been tested in animal models or in *in-vitro* system where the potential targets within the malarial parasites remain unknown [63].

Recently, novel approaches for the development of new drugs has been initiated by focusing on different modes of action of the drug (in comparison to existing ones) as well as more attention is laid toward the numerous metabolic and biochemical pathways of *plasmodium* parasites with the hope of identifying a novel pathway(s) or drug(s) against specific targets [60]. With the help of computational biology and network analysis, screening of various large compound libraries have made the way easy in short-listing the potential drug candidates for the targets of novel pathway(s) of malaria parasites [62,64,65]. In addition, advancement in mass spectrometry [66] methods has emerged the field of metabolomics [67] that is important in identifying and discovering the novel drug targets. It has helped in understanding the various metabolites and their responses toward the antimalarial drugs [68].

According to Harlow-Knapp effect, there are several novel pathways that have been identified for the targets and some of them are mentioned briefly here as it is currently an ongoing process.

5.6.1 The powerhouse: mitochondria

Mitochondria is called as the powerhouse of the cell; however, mitochondria of the *Plasmodium* is distinct in terms of several functions (biochemical and molecular) and also it plays a vital role in its life cycle [69]. It maintains the inner mitochondrial potential and helps in the development and growth of parasites. Recently, several members of the electron transport chain (ETC) have been identified as a novel target. The plasmodial mitochondrial ETC is the complex of non-proton and proton motive compounds/enzymes along-with ubiquinone and cytochrome $c1$ used as an electron carrier between the complexes [69]. The enzymes that are found to be important in novel targets are listed below :

Dihydroorotate dehydrogenase (DHODH): It plays a vital role in the oxidation of dihydroorotate to orotate, a precursor for the pyrimidine synthesis having the connection with ETC. DSM265 is reported as the first inhibitor of *Pf*DHODH and it is in the clinical trials.

Cytochrome *bc*1 (complex III): It is the only enzyme complex common to all respiratory ETCs and consists of 11 different polypeptides. It transfers the protons into intermembrane space through the redox reaction of ubiquinone in the Q cycle. Atovaquone is found to be the potential inhibitor of complex III; however, further validation is ongoing.

Type II NADH dehydrogenase (NDH2), mitochondrial glycerol-3-phosphate dehydrogenase (mG3DH), succinate dehydrogenase (SDH), malate-quinone oxyreductase (MQO), and ATPase are some of the other novel targets of ETCs of *Plasmodium* mitochondria about which Lunev *et al.* have described in detail [69].

5.6.2 The tricarboxylic acid cycle (TCA) and pyrimidine biosynthesis pathways

During the asexual stage of parasites, TCA metabolism comes across as an important metabolic pathway. However; they generally do rely on glycolysis during the blood stage of the life cycle. The exact function of the TCA is not fully understood, but there are several enzymes of TCA that are considered as targets for antimalarial drug development.

Aspartate aminotransferase (AspAT), malate dehydrogenase (MDH), and fumarate hydratase (FH) are currently under investigation for their antimalarial properties [69].

The crucial component is the synthesis of pyrimidine which helps the malarial parasite to spread into a human host via rapid replication of DNA [70]. It has been confirmed that human cells do pyrimidine synthesis through *de novo* or by salvaging pathway; whereas, parasite lacks the salvage pathways and depends only on *de novo* pathway for the synthesis as well as for their survival [71].

Therefore, **carbamoyl phosphate synthetase II (CPSII), aspartate transcarbamoylase (ATC), dihydroorotase, orotate phosphoribosyl transferase (OPRT), and orotidine 5′-monophosphate decarboxylase (OPDC)** are the promising targets, as ribozyme approach is much more target specific [72].

Besides the previously mentioned novel targets, an organ similar to the chloroplast (in plants) has been found in *Plasmodium* species called as **apicoplast**, a four membrane plastid organelle formed due to secondary endosymbiosis [73]. The genome sequencing data reveal the importance of the apicoplast as several metabolic pathways occurring inside the apicoplast are distinct from the humans, and they may be the potential drug targets. Malarial parasites mainly depend on type II fatty acid synthase (FAS II) in apicoplast for fatty acid biosynthesis due to lack of type I FAS, which is generally present in other eukaryotes, and make it the most favorable targets [74]. The FAS II further catalyzed by four enzymes named as **beta-ketoacyl ACP synthase ИI (FabB/F), b-ketoacyl-ACP reductase (FabG), enoyl-ACP reductase (Fab I), and b-hydroacyl-ACP dehydratase (FabZ)** that are equal holds an importance in the novel target identification [74].

Recent studies have also established that the nuclear-encoded proteins that are targeted to the apicoplast in *P. falciparum* due to the presence of N-terminal signal peptide sequence followed by a transit peptide sequence (which are usually not conserved) and their interacting partners such as *P. falciparum* Hsp 70-1 are potential new drug targets. Any drug molecule/peptide that inhibits the interaction between transit peptide-containing nuclear-encoded proteins and their interacting partners that assist their transport to apicoplast can be potential antimalarials which can disrupt the apicoplast metabolism important for parasite survival [75,76].

The postgenomic era has emerged and developed rapidly in the field of malarial biology filling the gap between host—pathogen interactions and the underlying molecular mechanism(s). The less studied and undercover area of complex systems are being explored opening a new window for the development of novel, target-specific, and efficient antimalarial through screening of the thousand potential targets in a quick span of time.

5.7 Conclusion

Among the numerous infectious diseases, the burden and death due to malaria is highest as reported by WHO, mostly in Africa followed by Asian countries. The developments in the field of epidemiology and entomology supported us not only in reducing the burden but also help in understanding the malarial parasite more rapidly. Numerous biological and human factors mediate the parasite and vector growth enhancing the disease burden. Although there have been antimalarial drug(s) available for the treatment of human malaria infection, continuous drug administration to clear *P. falciparum* allows parasites to sustain drug pressure and resist their therapeutic effect. The evolution of resistance against low-cost drugs finds an enormous societal cost for combating the spread of the disease. Vaccine development against malaria has just opened up with the development of RTS,S and needs more effective outputs. The recent development of proteomics and genomics has laid the stone of a new field that facilitates the identification of novel antimalarial targets with broader aspects and, hopefully, it will give better results.

Acknowledgments

Dr. Rajeev Tyagi would like to thank Dr. Pierre Druilhe from Institute Pasteur, Paris, France for allowing him to take inputs from his doctoral thesis. Sincere gratitude to Dr. John Adams, University of South Florida and Dr. Christopher Cutler, August University, USA for their support and guidance during his postdoctoral training. Dr. Tyagi would like to offer his sincere thanks to the Department of Science and Technology (DST), Govt. of India for funding part of this study (grant number: ECR/2015/000264). Mr. Nikunj Tandel thanks the Council of Scientific and Industrial Research (CSIR), New Delhi, Govt. of India for providing fellowship for his research (CSIR-SRF HRDG No.: 09/1048(0009)/2018-EMR-I).

References

[1] Nii-Trebi NI. Emerging and neglected infectious diseases: insights, advances, and challenges. BioMed Research International 2017;2017:15.

[2] Tibayrenc M. Encyclopedia of infectious diseases: modern methodologies. John Wiley & Sons; 2007.

[3] Cowman AF, Healer J, Marapana D, Marsh K. Malaria: biology and disease. Cell 2016; 167(3):610−24.

[4] WHO. World malaria report 2018. World Health Organization; 2018.

[5] Tyagi RK, Tandel N, Deshpande R, Engelman RW, Patel SD, Tyagi P. Humanized mice are instrumental to the study of *Plasmodium falciparum* infection. Frontiers in Immunology 2018;9(2550).

[6] Cox FE. History of the discovery of the malaria parasites and their vectors. Parasites and Vectors 2010;3(1):5.

[7] WHO. Global technical strategy for malaria 2016-2030. World Health Organization; 2015.

[8] WHO. Status report on artemisinin and ACT resistance. April 2017. WHO (WHO/HTM/GMP/2017.9). 2017:11.

[9] WHO. Global report on insecticide resistance in malaria vectors: 2010−2016. 2018.

[10] Ouji M, Augereau J-M, Paloque L, Benoit-Vical F. Plasmodium falciparum resistance to artemisinin-based combination therapies: a sword of Damocles in the path toward malaria elimination. Parasite 2018;25.

[11] Nsanzabana C. Resistance to artemisinin combination therapies (ACTs): do not forget the partner drug! Tropical Medicine and Infectious Disease 2019;4(1):26.

[12] Laurens MB. The promise of a malaria vaccine—are we closer? Annual Review of Microbiology 2018;72:273—92.

[13] Coelho CH, Doritchamou JYA, Zaidi I, Duffy PE. Advances in malaria vaccine development: report from the 2017 malaria vaccine symposium. NPJ Vaccines 2017;2(1):34.

[14] Alaganan A, Singh P, Chitnis CE. Molecular mechanisms that mediate invasion and egress of malaria parasites from red blood cells. Current Opinion in Hematology 2017;24(3):208—14.

[15] Ménard R, Tavares J, Cockburn I, Markus M, Zavala F, Amino R. Looking under the skin: the first steps in malarial infection and immunity. Nature Reviews Microbiology 2013;11(10):701—12.

[16] Phillips MA, Burrows JN, Manyando C, van Huijsduijnen RH, Van Voorhis WC, Wells TNC. Malaria. Nature Reviews, Diseases Primers 2017;3 (2056-676X (Electronic)).

[17] Bertolino P, Bowen DG. Malaria and the liver: immunological hide-and-seek or subversion of immunity from within? Frontiers in Microbiology 2015;6:41.

[18] Aly AS, Vaughan AM, Kappe SH. Malaria parasite development in the mosquito and infection of the mammalian host. Annual Review of Microbiology 2009;63:195—221.

[19] Dhangadamajhi G, Kar SK, Ranjit M. The survival strategies of malaria parasite in the red blood cell and host cell polymorphisms. Malaria Research and Treatment 2010; 2010.

[20] Lee AH, Symington LS, Fidock DA. DNA repair mechanisms and their biological roles in the malaria parasite *Plasmodium falciparum*. Microbiology and Molecular Biology Reviews 2014;78(3):469—86.

[21] Hugo RLE, Birrell GW. Proteomics of Anopheles vectors of malaria. Trends in Parasitology 2018;34(11):961—81.

[22] Tangpukdee N, Duangdee C, Wilairatana P, Krudsood S. Malaria diagnosis: a brief review. Korean Journal of Parasitology 2009;47(2):93.

[23] Krampa F, Aniweh Y, Awandare G, Kanyong P. Recent progress in the development of diagnostic tests for malaria. Diagnostics 2017;7(3):54.

[24] Ray S, Reddy PJ, Jain R, Gollapalli K, Moiyadi A, Srivastava S. Proteomic technologies for the identification of disease biomarkers in serum: advances and challenges ahead. Proteomics 2011;11(11):2139—61.

[25] Verma P, Biswas S, Mohan T, Ali S, Rao D. Detection of histidine rich protein & lactate dehydrogenase of *Plasmodium falciparum* in malaria patients by sandwich ELISA using in-house reagents. Indian Journal of Medical Research 2013;138(6):977.

[26] Zheng Z, Cheng Z. Advances in molecular diagnosis of malaria. Advances in Clinical Chemistry 2017;80:155—92.

[27] Foxman B. Molecular tools and infectious disease epidemiology. Academic Press; 2010.

[28] Wilson NO, Adjei AA, Anderson W, Baidoo S, Stiles JK. Detection of *Plasmodium falciparum* histidine-rich protein II in saliva of malaria patients. The American Journal of Tropical Medicine and Hygiene 2008;78(5):733—5.

[29] Pooe OJ. The detection of *Plasmodium falciparum* in human saliva samples. 2011.

[30] Fung AO, Damoiseaux R, Grundeen S, et al. Quantitative detection of Pf HRP2 in saliva of malaria patients in the Philippines. Malaria Journal 2012;11(1):175.

[31] Tao D, McGill B, Hamerly T, et al. A saliva-based rapid test to quantify the infectious subclinical malaria parasite reservoir. Science Translational Medicine 2019;11(473).

[32] Oguonu T, Shu E, Ezeonwu BU, et al. The performance evaluation of a urine malaria test (UMT) kit for the diagnosis of malaria in individuals with fever in south-east Nigeria: cross-sectional analytical study. Malaria Journal 2014;13(1):403.

[33] Putaporntip C, Buppan P, Jongwutiwes S. Improved performance with saliva and urine as alternative DNA sources for malaria diagnosis by mitochondrial DNA-based PCR assays. Clinical Microbiology and Infections 2011;17(10):1484—91.

[34] Ghayour Najafabadi Z, Oormazdi H, Akhlaghi L, et al. Detection of *Plasmodium vivax* and *Plasmodium falciparum* DNA in human saliva and urine: loop-mediated isothermal amplification for malaria diagnosis. Acta Tropica 2014;136:44—9.

[35] AE SM, El-Rayah el-A, Giha HA. Towards a noninvasive approach to malaria diagnosis: detection of parasite DNA in body secretions and surface mucosa. Journal of Molecular Microbiology and Biotechnology 2010;18(3):148—555.

[36] Ashley EA, Pyae Phyo A, Woodrow CJ. Malaria. The Lancet 2018;391(10130): 1608—21.

[37] Sinka ME, Bangs MJ, Manguin S, et al. A global map of dominant malaria vectors. Parasites and Vectors 2012;5(1):69.

[38] Killeen GF. Characterizing, controlling and eliminating residual malaria transmission. Malaria Journal 2014;13(1):330.

[39] Protopopoff N, Van Bortel W, Speybroeck N, et al. Ranking malaria risk factors to guide malaria control efforts in African highlands. PLoS One 2009;4(11):e8022.

[40] Bødker R, Msangeni HA, Kisinza W, Lindsay SW. Relationship between the intensity of exposure to malaria parasites and infection in the Usambara Mountains, Tanzania. The American Journal of Tropical Medicine and Hygiene 2006;74(5):716—23.

[41] Caminade C, Kovats S, Rocklov J, et al. Impact of climate change on global malaria distribution. Proceedings of the National Academy of Sciences United States of America 2014;111(9):3286—91.

[42] Kar NP, Kumar A, Singh OP, Carlton JM, Nanda N. A review of malaria transmission dynamics in forest ecosystems. Parasites and Vectors 2014;7(1):265.

[43] Hemingway J, Shretta R, Wells TN, et al. Tools and strategies for malaria control and elimination: what do we need to achieve a grand convergence in malaria? PLoS Biology 2016;14(3):e1002380.

[44] WHO. Malaria: fact sheet. World Health Organization. Regional Office for the Eastern Mediterranean; 2017.

[45] WHO. WHO-date accessed malaria vaccine rainbow tables. World Health Organization; 2017.

[46] Dhiman S. Are malaria elimination efforts on right track? An analysis of gains achieved and challenges ahead. Infectious Diseases of Poverty 2019;8(1):14.

[47] WHO. Malaria: fact sheet. World Health Organization. Regional Office for the Eastern Mediterranean; 2019.

[48] Kaehler N, Adhikari B, Cheah PY, et al. The promise, problems and pitfalls of mass drug administration for malaria elimination: a qualitative study with scientists and policymakers. International Health 2019;11(3):166—76.

[49] WHO. Guidelines for the treatment of malaria. World Health Organization; 2015.

[50] Hetzel MW, Genton B. Mass drug administration for malaria elimination: do we understand the settings well enough? BMC Medicine 2018;16(1):239.

[51] Von Seidlein L, Dondorp A. Fighting fire with fire: mass antimalarial drug administrations in an era of antimalarial resistance. Expert Review of Anti-infective Therapy 2015;13(6):715−30.

[52] Grueninger H, Hamed K. Transitioning from malaria control to elimination: the vital role of ACTs. Trends in Parasitology 2013;29(2):60−4.

[53] Newby G, Hwang J, Koita K, et al. Review of mass drug administration for malaria and its operational challenges. The American Journal of Tropical Medicine and Hygiene 2015;93(1):125−34.

[54] Brady OJ, Slater HC, Pemberton-Ross P, et al. Role of mass drug administration in elimination of *Plasmodium falciparum* malaria: a consensus modelling study. The Lancet Global Health 2017;5(7):e680−7.

[55] Morris U, Msellem MI, Mkali H, et al. A cluster randomised controlled trial of two rounds of mass drug administration in Zanzibar, a malaria pre-elimination setting—high coverage and safety, but no significant impact on transmission. BMC Medicine 2018; 16(1):215.

[56] von Seidlein L, Peto TJ, Landier J, et al. The impact of targeted malaria elimination with mass drug administrations on falciparum malaria in Southeast Asia: a cluster randomised trial. PLoS Medicine 2019;16(2):e1002745.

[57] Elimination mRCPoTfM. malERA: an updated research agenda for diagnostics, drugs, vaccines, and vector control in malaria elimination and eradication. PLoS Medicine 2017;14(11):e1002455.

[58] Alonso PL, Brown G, Arevalo-Herrera M, et al. A research agenda to underpin malaria eradication. PLoS Medicine 2011;8(1):e1000406.

[59] Adepoju P. RTS, S malaria vaccine pilots in three African countries. The Lancet 2019; 393(10182):1685.

[60] Kumar S, Bhardwaj T, Prasad D, Singh RK. Drug targets for resistant malaria: historic to future perspectives. Biomedicine and Pharmacotherapy 2018;104:8−27.

[61] Alam A, Goyal M, Iqbal MS, et al. Novel antimalarial drug targets: hope for new antimalarial drugs. Expert Review of Clinical Pharmacology 2009;2(5):469−89.

[62] Santos G, Torres NV. New targets for drug discovery against malaria. PLoS One 2013; 8(3):e59968.

[63] Antony HA, Parija SC. Antimalarial drug resistance: an overview. Tropical Parasitology 2016;6(1):30.

[64] Ludin P, Woodcroft B, Ralph SA, Mäser P. In silico prediction of antimalarial drug target candidates. International Journal for Parasitology: Drugs and Drug Resistance 2012;2:191−9.

[65] Tran T, Ekenna C. Metabolic pathway and graph identification of new potential drug targets for *Plasmodium Falciparum*. In: Paper presented at: 2017 IEEE international conference on bioinformatics and biomedicine (BIBM); 2017.

[66] Misra G, editor. Data processing handbook for complex biological data sources. Academic Press; March 23, 2019.

[67] Arivaradarajan P, Misra G, editors. Omics approaches, technologies and applications: integrative approaches for understanding OMICS data. Springer; February 4, 2019.

[68] Allman EL, Painter HJ, Samra J, Carrasquilla M, Llinás M. Metabolomic profiling of the malaria box reveals antimalarial target pathways. Antimicrobial Agents and Chemotherapy 2016;60(11):6635−49.

[69] Lunev S, Batista FA, Bosch SS, Wrenger C, Groves MR. Identification and validation of novel drug targets for the treatment of *Plasmodium falciparum* malaria: new insights.

Current Topics in Malaria: Intech November 30, 2016:235–65. https://doi.org/10.5772/65659.

[70] White N, Pukrittayakamee S, Hien T, Faiz M, Mokuolu O, Dondorp A. Erratum: malaria (the lancet (2014) 383 (723-735). The Lancet 2014;(9918):383.

[71] Vaidya AB, Mather MW. Mitochondrial evolution and functions in malaria parasites. Annual Review of Microbiology 2009;63:249–67.

[72] Alvarez-Salas LM. Nucleic acids as therapeutic agents. Current Topics in Medicinal Chemistry 2008;8(15):1379–404.

[73] Wilson RI. Parasite plastids: approaching the endgame. Biological Reviews 2005;80(1): 129–53.

[74] Qidwai T, Khan F. Antimalarial drugs and drug targets specific to fatty acid metabolic pathway of *Plasmodium falciparum*. Chemical Biology and Drug Design 2012;80(2): 155–72.

[75] Misra G, Ramachandran R. Hsp70-1 from *Plasmodium falciparum*: protein stability, domain analysis and chaperone activity. Biophysical Chemistry June 1, 2009; 142(1–3):55–64.

[76] Misra G, Ramachandran R. Exploring the positional importance of aromatic residues and lysine in the interactions of peptides with the *Plasmodium falciparum* Hsp70-1. Biochimica et Biophysica Acta November 1, 2010;1804(11):2146–52.

Sleeping sickness

6

Dusit Laohasinnarong

Assistant Professor, Department of Clinical Sciences and Public Health, Faculty of Veterinary Science, Mahidol University, Salaya, Nakhon Pathom, Thailand

6.1 Introduction

Sleeping sickness or human African trypanosomosis (HAT) is a vector-borne disease majorly prevalent in sub-Saharan Africa [1,2]. It is named after its major symptom, sleep—wake cycle. This disease is a neglected tropical disease responsible for the top three deaths of the African people. It is caused by infection with a flagellated protozoan in the family of Trypanosomatidae, the genus *Trypanosoma* (taxonomy is shown in Fig. 6.1). The causal agent is the member of subgenus *Trypanozoon*, subspecies of *Trypanosoma brucei*: *Trypanosoma brucei rhodesiense* and *Trypanosoma brucei gambiense*. Sleeping sickness can be divided by geographical spreading domain, namely Eastern and Western sleeping sickness. The pathogens are mechanically spread by tsetse, bloody insect of the genus *Glossina*. They are also able to infect other mammalian hosts that act as a reservoir, for example, cattle, goats, and wild mammals [3—5].

Kingdom	Protista
Subkingdom	Protozoa
Phylum	Sarcomastigophora
Subphylum	Mastigophora
Class	Zoomastigophora
Order	Kinetoplastida
Family	Trypnosomatidae
Section	Salivaria
Genus	*Trypanosoma*
Subgenus	*Trypanozoon*
Species	*Brucei*

FIGURE 6.1

Taxonomy of trypanosome.

Molecular Advancements in Tropical Diseases Drug Discovery. https://doi.org/10.1016/B978-0-12-821202-8.00006-2

Eastern or Rhodesian sleeping sickness is an acute form that occurs in the eastern and southern parts of Africa. The major symptoms can be detected within a week after infection. The death occurs less than 6 months of infection. The disease is caused by the infection of *T. b. rhodesiense* in which *Glossina morsitans morsitans* is an important vector. It has quite a complex zoonosis because domestic and wild animals are reservoir hosts. Thereby, Eastern sleeping sickness is a zoonosis [6−8].

Western or Gambian sleeping sickness shows the chronic form and is found in the west and central Africa. The infected persons may not show any symptoms for months or years. Death may occur several years after the onset of the disease. The disease is caused by *T. b. gambiense* that has *G. palpalis* as the significant vector in this region. The human-to-human infection cycle can occur for Western HAT [9−12].

Sleeping sickness, both forms, is noxious if the treatment is too late or remains untreated. However, the drug for the treatment of the end stage is relatively toxic. Nonetheless, due to the decrease in the number of new cases since 2000, the World Health Organization aimed to remove this disease from public health problems in 2020.

6.2 Detection of infection

Nowadays, some diseases have similar clinical signs as sleeping sickness, causing difficulty in demarcation. Therefore laboratory diagnosis plays an extremely important role. Even so, the required tests are different based on the form of the disease. For the late stage, high sensitivity and specificity of the technique are very essential to avoid the development of resistance toward trypanocidal medicines. Many tests based on the available technologies have been developed and applied in the detection of trypanosomes [13−18]. However, the most required features of diagnostics are rapid, simple, reliable, and applicable in the field. In addition, the development of new diagnostics is necessary to keep up with changes or mutations in the parasite.

Parasites detection using microscopy is the first method that is used to diagnose the infection. It is used as a gold standard technique that detects the parasite directly from the specimen [19,20]. Basically, blood smear is good to detect trypanosomes because the parasite is live in blood and tissue. The movement of trypanosomes among the red blood cells can be detected in fresh blood spot under the coverslip. Hematocrit centrifuge technique (HCT) has been introduced to detect trypanosomes with higher sensitivity than blood film [21]. The trypanosomes can be found in the buffy coat area under a microscope. This technique can enhance the sensitivity compared to the direct blood smear. However, HCT is mostly applied in animal trypanosomosis, not HAT. To detect trypanosomes in sleeping sickness, the miniature anion-exchange centrifugation technique (m-AECT) is mostly applicable [22,23]. m-AECT can separate the trypanosomes from patient blood through anion-exchange column. This column contains diethyl amino-ethyl cellulose type DE 52, which allows only parasites to pass through this cellulose. Even the original m-AECT is designed to be used in the field, some new versions simplified the technique as cost-effective and modified for its use in the lab [24]. m-AECT does not take much time and the collected parasites can be identified. The parasite detection technique is the fundamental method that is easy, simple to practice and apply in the remote areas. It has high specificity but low sensitivity. Likewise, it depends on the level of parasitemia, trypanosome species, and the laboratory technician's experience. It is beneficial for a definite diagnosis, especially the late state, but not for an epidemiological study.

Serological methods have also applied as diagnostic tools for both antigen and antibody detection [25,26]. In general, west African sleeping sickness uses the card agglutination test for trypanosomiasis (CATT) as the basic screening test followed by confirmation of the infection by parasite detection. This diagnosis procedure cannot apply for rhodesiense HAT because the infected persons will show the clinical signs as early as a week after infection. The doctor can diagnose by the symptoms. The serological test has a limitation as the detection of antibodies is unclear, as the causing agent may be completely eliminated but the level of immunity remains. So, the confirmation of infection is based on the detection of antigen.

Currently, molecular techniques are used with lower cost thus enabling lower-middle-income countries also to procure the system. Nucleotide amplification methods have proved as alternative assays with higher sensitivity and specificity than parasite detection and serological tests [27−29]. Polymerase chain reaction (PCR) is the most favorite molecular method and has been developed for the detection of trypanosome DNA. PCR application for diagnostic sleeping sickness can be divided into two types: One that detects subgenus and other is species specific. Subgenus *Trypanozoon* has many identical genes, which makes the development of species-specific tests much more complicated.

The isothermal nucleic acid amplification assays are a simpler option. Loop-mediated isothermal amplification (LAMP) is widely applied to detect several pathogens, including trypanosomes [30]. LAMP can detect both DNA and RNA, simple to perform, and does not need any special instrument [31,32]. The technique uses less reaction time with higher sensitivity and specificity compared to traditional PCR [5,33]. For economic reasons, the vast majority of similar genomes in subgenus *Trypanozoon*, and geographic disease separation makes the development of *Trypanozoon* or *T. brucei* test kit, a good solution. LAMP assay for HAT, however, has been developed for both subgenus *Trypanozoon* and subspecies *T. brucei s.l.* [14,16,34,35]. For the species specific, the development is based on the existing specific target gene for that species, that is *T. b. gambiense* specific glycoprotein (TgsGP) and Serum resistance-associated gene. Many developments in PCR and LAMP for subgenus *Trypanozoon* detection are based on several genes such as internal transcribed sequence, *para*-flagella rod A, and repetitive insertion mobile element [14,34,36,37]. *Trypanozoon* LAMP has shown higher sensitivity and specificity than traditional PCR. So, this technique was the choice for field samples as the new alternative method for HAT diagnosis. Since 2011, the LAMP test trial in the field for sleeping sickness was done collaboratively between FIND and Eiken Chemical [38].

The diagnosis of east HAT is not so complicated as compared to west HAT. *T. b. gambiense* sleeping sickness diagnosis will start using CATT. Then the positive sample will be sent for parasite detection by m-AECT for confirmation. Even m-AECT is sensitive to detect low parasitemia of less than 50 trypanosomes/mL; however, there is no guarantee of the false negative in the very low parasitemia cases, for example, <10 parasites/mL. Nowadays, the prevalence of HAT is low with a small number of new cases. Monitoring and surveillance with the use of highly accurate test kits is essential. LAMP technique has shown that it can detect as low as 1 parasite/mL [15,16,28,34,35,39]. Thus LAMP may be applied for the second-step diagnosis of gambiense sleeping sickness, instead of m-AECT. Furthermore, LAMP can detect unprocessed samples and uses less blood volume of only 2 μL in case of direct detection. LAMP reagent can be stored in the refrigerator or at a temperature lower than the recommendation (25−37°C) when needed for fieldwork [39]. Both PCR and LAMP for HAT have been evaluated for the second-stage sleeping sickness with cerebrospinal fluid samples. The tests, however, cannot predict the progress of disease or treatment.

6.3 Epidemiology and risk factors

Approximately 36 countries in sub-Saharan Africa, located at latitude 14°N and 20°S, are vulnerable to sleeping sickness. Western HAT occurs in 24 countries, while eastern HAT is found in 13 countries. Only Uganda has both forms of sleeping sickness, but there are no concurrent epidemic areas [40]. The disease spreads mainly in rural areas. The number of cases is mostly (approximately 98%) gambiense HAT. The number of reported cases has dropped to below 10,000 since 2009. Presently, the reported cases are less than 2000 cases annually. Nevertheless, the majority of epidemic areas still remain the same, the Democratic Republic of the Congo [29].

Sleeping sickness is mechanically transmitted by tsetse, a blood-biting vector. The life cycle of trypanosome is divided into two main stages, in vector phase and in the hosts phase (Fig. 6.2). There are various habitats of tsetse responsible for the different transmission cycles. *T. b. gambiense* sleeping sickness is transmitted by *Palpalis* group or riverine files (subgenus *Palpalis* or formerly named *Nemorhina*) of tsetse, *G. palpalis*, *G. p. gambiensis*, and *G. fuscipes* [10,41]. On the other hand, rhodesinse HAT is transmitted by *Morsitans* group or Savannah files (subgenus *Glossina* s.s.), for example, *G. morsitans*, *G. swynnertoni*, and *G. pllidipes* [10,40]. The habitat of major vectors for Rhodesian HAT is forest while the vectors of Gambian HAT are near the river. The presence of tsetse, however, may not always indicate this disease, which remains unexplainable. West African sleeping sickness has an anthroponotic cycle but eastern African sleeping sickness is zoonotic as shown in Fig. 6.3.

6.4 Approaches to control and elimination through mass drug administration

African sleeping sickness is a vector-borne disease; the two levels for disease control are pathogen and vector control. The vector control is considered after trying to

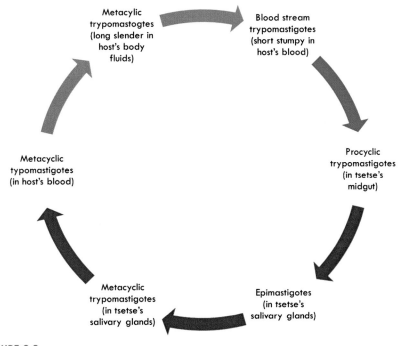

FIGURE 6.2

Life cycle of trypanosome.

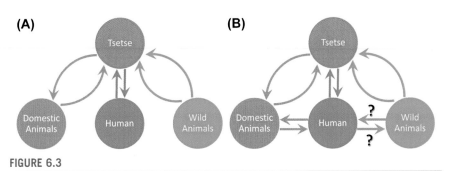

FIGURE 6.3

Transmission cycle of (A) Rhodesian human African trypanosomosis and (B) Gambian human African trypanosomosis

control the infection in humans. With the reduction of new cases, however, the control of vectors may not be that important and the vaccine will be a better option to control the disease. Samia et al. [42] evaluated using a mathematical model that the use of vaccines for disease prevention is not enough, and it must be coupled with the control of tsetse. In addition, the control of insect vectors by using insecticides is better than using vaccines. The vaccine development, however, is not quite easy because trypanosomes have variant surface glycoprotein (VSG). The parasites avail this antigenic variation against the host's antibody response [43,44].

Table 6.1 Sleeping sickness or human African trypanosomosis (HAT).

	Rhodesian sleeping sickness	Gambian sleeping sickness
Spread area	Eastern and Southern Africa	Western and Central Africa
Form	Acute	Chronic
Causative agent	*Trypanosoma brucci rhodesiense*	*Trypanosoma brucei gambiense*
Vector	*Glossina morsitan morsitan*	*Glossina palpalis*
Transmission route	Sexual, congenital	
Transmission cycle	Human-tsetse-human Human-tsetse-wild/domestic animals-tsetse-human	Human-tsetse-human
Reservoirs	Domestic and wild animals	?
Zoonosis	Yes	?
Drugs for early stage	Suramin	Pentamidine, suramin
Drugs for late stage	Melarsoprol	Melarsoprol, eflornithine, nifurtimox-eflornithine
Vaccine	Not available	Not available

The use of antitrypanosomal drugs depends on the stage of the disease. The available trypanocides for the first and second phases of HAT are shown in Table 6.1, which have only five useable antitrypanosomal drugs, suramin, pentamindine, melarsoprol, eflornithine, and nifurtimox-eflornithine combination [17,45,46]. The first two drugs are used for the first stage while others are used in the late stage. The rising concern of drug resistance in the presently used trypanocidal drugs was reported [47−51]. Definite diagnosis is crucial before the decision on the drug regimen is taken.

Suramin was introduced in 1920 against *T. b. rhodesiense* infection. *T. b. gambiense* infection is generally cured using pentamidine. Both trypanocides can be applied only in the early stage, and these drugs do not pass through the blood−brain barrier [52−55]. The adverse effects, including anaphylactic shock, renal failure, and neurotoxicity, may occur in suramin more than pentamidine treatment. For the late stage, melarsoprol, launched in 1949, can cure the Rhodesian sleeping sickness, while eflornithine, approved by US FDA in 1990, can be used for *T. b. gambiense* infection [45,56,57]. Melarsoprol is arsenic-based trypanocide and has a higher toxicity than eflornithine. Drug resistance is a critical issue because there are very few available trypanocidal medicines. The concept of applying the existing chemotherapy of Chaga's disease for the treatment of sleeping sickness has taken and found that nifurtimox-eflornithine combination therapy (NECT) can treat Gambian trypanosomosis [46,58]. In the second phase, the parasites invade and pass blood−brain barrier into the nervous system. The late-stage trypanocidal drugs have side effects and may result in death. Therefore, the development of antitrypanosomal drugs for the treatment of the second stage is very important.

The control strategy is different by the forms of sleeping sickness. Rhodesiense sleeping sickness is more complex than gambiense HAT because domestic and wild animals are reservoir hosts. In the epidemic area, evacuation of people is a good strategy that helps to reduce the risk but may cause problems in arable land. Therefore, control of tsetse is the other feasible choice that can be used instead. Gambian sleeping sickness needs to be identified properly before starting treatment. The mass chemotherapy and chemo-prophylaxis may be applied in the gambiense HAT [59,60] when the knowledge about the stage of infection is missing with the absence of clinical signs. Mass chemotherapy and or trypanocidal prophylaxis are also applied in the reservoir hosts if the infection is prevalent in livestock or humans [60].

Beyond antitrypanosomal drugs, vector control is an approach to be selected. Insecticides are generally the first line that is used [61−63]. Mostly, insecticides employed to control tsetse are pyrethroid pesticides due to their low toxicity to mammals and fast degradation. They are organic compounds broadly used to control vectors [64]. Pesticides used to control tsetse include cypermethrin, deltamethrin, and flumethrin [65−70]. Dung fauna and dung beetles, however, get affected due to the use of insecticides by spraying on animals or in rearing areas [71]. In addition, the use of the same pesticide for a long time may lead tsetse to develop resistance. Thus limiting the exposure amount to the same areas or changing insecticide can help to reduce such problems. Fipronil was used against tick in Latin America. Its efficacy against tsetse in Africa was evaluated and found to be valuable and effective [70,72]. The use of natural enemies is another interesting option. Many insect pathogens were investigated to kill or sterile tsetse [73,74]. The control of trypanosome colonization by microorganisms, such as bacteria, is also a preferred method [75,76]. Studies on natural enemies of tsetse that aims to replace other methods that may have side effects on humans have been conducted.

The disease control has to include several means such as surveillance in vectors and hosts, accurate diagnostics, chemotherapy, and consideration of the local environment. The disease eradication is quite challenging because the disease forms are different in various geographical areas and the disease occurs in remote areas.

6.5 Next-generation vaccines, drugs, and diagnostics

Sleeping sickness is one of the major neglected tropical diseases that occur in sub-Saharan Africa only. The disease is found in low-income countries, so the investment in new drug development is not worthwhile. The available curative drugs for HAT have been used for almost a century. Although some medicines, especially for the second stage, are toxic, still they are used. Instead of focusing on research to find new chemotherapeutic drugs, the application of the combination therapy is practiced involving drugs used against Chaga's disease along with effective trypanocide, NECT. Drug resistance should be continuously monitored [50,77,78].

Immune evasion mechanism of trypanosomes by antigenic variation, VSGs gene, makes the vaccine development quite challenging [43,79,80]. VSGs have two terminal domains, N- (exposed to the immune system) and C- (conserved in the coat), in which about 7% are fully functional [81,82]. The reservoir host is a major target for a vaccine. Once we can control the vector and immunize the reservoir host, it is quite easy to control HAT. Universal protection by the vaccine is the goal of controlling HAT if we can use a conserved gene to produce effective vaccines. VSG is high antigenic and the host immune system responds very well, especially on the first infection. Thereafter the parasites will change their shape and the set of VSG, causing the antibody to not recognize these new parasites. VSG, therefore, is not a good candidate for vaccine development. The vaccine development research has changed the focus to the cytoskeleton. The cytoskeletal structure consists of many proteins such as microtubules, tubulin, actin, paraflagellar rod, and flagellum attachment zone [83,84]. Tubulin, microtubules, and actin have been used to develop vaccines. This protein from T. brucei has the potential to be used for vaccine development when compared to tubulin from T. evansi accounting for higher protection rate [85–89]. In addition, the adjuvant and route of injection are also important for protection [85,86].

Traditional medicine has a long history. Every country in the world has used traditional medicine, which may vary according to the area. The use of medicinal plants instead of chemicals has been preferred because it is safer and less toxicity [49,90,91]. The purpose of using herbs will be to treat the disease and eliminate insect vectors. Numerous studies have focused on the therapeutic use of medicinal plants [92–96]. Phytomedicines can be used as fed, crude extract or purified phytochemicals [93,97–101]. Many herbs have been studied to kill African trypanosomes such as turmeric, Chinese skullcap, Japanese goldthread, aromatic herb of Artemisia spp [92,94,100,102]. Herbs can be applied as an insect repellent, for example, the use of citronella to repel mosquitoes.

Based on the existing gene targets and pathways, there are chances of drug or vaccine discovery. It is interesting to apply the genes or pathways that are intensively studied in American trypanosomes but not in African trypanosomes such as trans-sialidase and glutathione (or trypanothione) [103–107]. Among these genes, trypanothione is the most study in African trypanosomes [108–110].

6.6 Concluding remarks

Sleeping sickness is a neglected tropical disease that mainly occurs in sub-Saharan Africa. The disease frequently occurs in remote regions that are poor in resources. Therefore, there is a need for rapid and simple test kits as a point-of-care test (POCT). The diagnostic techniques that can apply inexpensive devices and little maintenance are suitable for use in the distant parts. Western HAT is a chronic form that needs proactive control measures. Frequent surveys to find patients are a way to help cure the disease before it enters the late stage. Abandonment of infected persons without treatment will increase the reservoir. This is different from the *T. b. rhodesiense* caused infection where the domestic and wild animals are an important reservoir that must be monitored and controlled. Early detection of both forms of HAT is crucial to control its transmission and prompt treatment with nontoxic drugs.

Candidate chemotherapeutic drugs should be safe, inexpensive, and should be produced locally. Medicinal plants are better alternatives that need further study. Some proteins are proposed as drug or vaccine targets. Epigenetic and genetic engineering studies are essential for drug and vaccine. The control of tsetse by insecticide may affect the environment, dung fauna, and other useful insects or predators.

Acknowledgment

I wish to express my deepest gratitude to Prof. Dr. Noboru Inoue at the National Research Center for Protozoan Diseases (NRCPD), Obihiro University of Agriculture and Veterinary Medicine, who introduced me to African trypanosomes. African friends who help me develop an understanding of human African trypanosomosis.

References

[1] Maudlin I, Eisler MC, Welburn SC. Neglected and endemic zoonoses. Philosophical Transactions of the Royal Society of London B Biological Sciences September 27, 2009;364(1530):2777−87.

[2] Büscher P, Cecchi G, Jamonneau V, Priotto G. Human African trypanosomiasis. The Lancet November 25, 2017;390(10110):2397−409.

[3] Kaare MT, Picozzi K, Mlengeya T, Fèvre EM, Mellau LS, Mtambo MM, et al. Sleeping sickness–a re-emerging disease in the Serengeti? Travel Medicine and Infectious Disease March 2007;5(2):117−24.

[4] Mekata H, Konnai S, Simuunza M, Chembensofu M, Kano R, Witola WH, et al. Prevalence and source of trypanosome infections in field-captured vector flies (*Glossina pallidipes*) in southeastern Zambia. Journal of Veterinary Medical Science September 2008;70(9):923−8.

[5] Laohasinnarong D, Thekisoe OMM, Malele I, Namangala B, Ishii A, Goto Y, et al. Prevalence of *Trypanosoma* sp. in cattle from Tanzania estimated by conventional PCR and loop-mediated isothermal amplification (LAMP). Parasitology Research July 8, 2011;109(6):1735−9.

[6] Enyaru JCK, Matovu E, Nerima B, Akol M, Sebikali C. Detection of *T. b. rhodesiense* trypanosomes in humans and domestic animals in south east Uganda by amplification of serum resistance-associated gene. Annals of the New York Academy of Sciences October 2006;1081:311−9.

[7] Laohasinnarong D, Goto Y, Asada M, Nakao R, Hayashida K, Kajino K, et al. Studies of trypanosomiasis in the Luangwa valley, north-eastern Zambia. Parasites and Vectors September 30, 2015;8(1):497.

[8] Matthews KR. 25 years of African trypanosome research: from description to molecular dissection and new drug discovery. Molecular and Biochemical Parasitology March 2015;200(1−2):30−40.

[9] Kuzoe FAS. Current situation of African trypanosomiasis. Acta Tropica September 1, 1993;54(3):153—62.

[10] Welburn SC, Fèvre EM, Coleman PG, Odiit M, Maudlin I. Sleeping sickness: a tale of two diseases. Trends in Parasitology January 1, 2001;17(1):19—24.

[11] Solano P, Jamonneau V, N'Guessan P, N'Dri L, Dje NN, Miezan TW, et al. Comparison of different DNA preparation protocols for PCR diagnosis of Human African Trypanosomosis in Côte d'Ivoire. Acta Tropica June 2002;82(3):349—56.

[12] Lejon V, Ngoyi DM, Kestens L, Boel L, Barbé B, Betu VK, et al. Gambiense human African trypanosomiasis and immunological memory: effect on phenotypic lymphocyte profiles and humoral immunity. PLoS Pathogens March 6, 2014;10(3):e1003947.

[13] Woo PT. Evaluation of the haematocrit centrifuge and other techniques for the field diagnosis of human trypanosomiasis and filariasis. Acta Tropica 1971;28(3):298—303.

[14] Njiru ZK, Mikosza ASJ, Matovu E, Enyaru JCK, Ouma JO, Kibona SN, et al. African trypanosomiasis: sensitive and rapid detection of the sub-genus *Trypanozoon* by loop-mediated isothermal amplification (LAMP) of parasite DNA. International Journal for Parasitology April 2008;38(5):589—99.

[15] Njiru ZK, Mikosza ASJ, Armstrong T, Enyaru JC, Ndung'u JM, Thompson ARC. Loop-mediated isothermal amplification (LAMP) method for rapid detection of *Trypanosoma brucei rhodesiense*. PLoS Neglected Tropical Diseases February 6, 2008;2(2): e147.

[16] Njiru ZK, Traub R, Ouma JO, Enyaru JC, Matovu E. Detection of group 1 *Trypanosoma brucei gambiense* by loop-mediated isothermal amplification. Journal of Clinical Microbiology April 1, 2011;49(4):1530—6.

[17] Kennedy PG. Clinical features, diagnosis, and treatment of human African trypanosomiasis (sleeping sickness). The Lancet Neurology February 1, 2013;12(2):186—94.

[18] Nambala P, Musaya J, Hayashida K, Maganga E, Senga E, Kamoto K, et al. Comparative evaluation of dry and liquid RIME LAMP in detecting trypanosomes in dead tsetse flies. Onderstepoort Journal of Veterinary Research October 3, 2018;85(1): a1543.

[19] Murray M, Murray PK, McIntyre WIM. An improved parasitological technique for the diagnosis of African trypanosomiasis. Transactions of the Royal Society of Tropical Medicine and Hygiene 1977;71(4):325—6.

[20] Truc P, Bailey JW, Doua F, Laveissière C, Godfrey DG. A comparison of parasitological methods for the diagnosis of gambian trypanosomiasis in an area of low endemicity in Côte d'Ivoire. Transactions of the Royal Society of Tropical Medicine and Hygiene July 1, 1994;88(4):419—21.

[21] Woo PTK. The haematocrit centrifuge for the detection of trypanosomes in blood. Canadian Journal of Zoology September 1, 1969;47(5):921—3.

[22] Lumsden WHR, Kimber CD, Strange M. *Trypanosoma brucei*: detection of low parasitaemias in mice by a miniature anion-exchanger/centrifugation technique. Transactions of the Royal Society of Tropical Medicine and Hygiene January 1, 1977; 71(5):421—4.

[23] Truc P, Jamonneau V, N'Guessan P, Diallo PB, Garcia A. Parasitological diagnosis of human African trypanosomiasis: a comparison of the QBC® and miniature anion-exchange centrifugation techniques. Transactions of the Royal Society of Tropical Medicine and Hygiene May 1, 1998;92(3):288—9.

[24] Truc P, Jamonneau V, N'Guessan P, N'Dri L, Diallo PB, Butigieg X. Simplification of the miniature anion exchange centrifugation technique for the parasitological

diagnosis of human African trypanosomiasis. Transactions of the Royal Society of Tropical Medicine and Hygiene September 1, 1998;92(5). 512−512.

[25] Truc P, Lejon V, Magnus E, Jamonneau V, Nangouma A, Verloo D, et al. Evaluation of the micro-CATT, CATT/Trypanosoma brucei gambiense, and LATEX/T b gambiense methods for serodiagnosis and surveillance of human African trypanosomiasis in West and Central Africa. Bulletin of the World Health Organization 2002;80(11):882−6.

[26] Chappuis F, Loutan L, Simarro P, Lejon V, Büscher P. Options for field diagnosis of human African trypanosomiasis. Clinical Microbiology Reviews January 1, 2005; 18(1):133−46.

[27] Cox A, Tilley A, McOdimba F, Fyfe J, Eisler M, Hide G, et al. A PCR based assay for detection and differentiation of African trypanosome species in blood. Experimental Parasitology September 2005;111(1):24−9.

[28] Chimbevo LM, Malala JB, Oshule PS, Muchiri WF, Otundo DO, Essuman S, et al. A comparison of parasitological techniques (microscopy) and loop-mediated isothermal amplification (LAMP) of DNA in diagnosis and monitoring treatment of Trypanosoma brucei rhodesiense infection in vervet monkeys (Chlorocebus aethiops). International Journal of Biochemistry Research and Review August 1, 2017;18(2):1−14.

[29] Boelaert M, Mukendi D, Bottieau E, Kalo Lilo JR, Verdonck K, Minikulu L, et al. A phase III diagnostic accuracy study of a rapid diagnostic test for diagnosis of second-stage human African trypanosomiasis in the Democratic Republic of the Congo. EBioMedicine January 1, 2018;27:11−7.

[30] Laohasinnarong D. Loop-mediated isothermal amplification (LAMP): an alternative molecular diagnosis. Journal of Applied Animal Science September 2011;4(3): 9−19.

[31] Notomi T, Okayama H, Masubuchi H, Yonekawa T, Watanabe K, Amino N, et al. Loop-mediated isothermal amplification of DNA. Nucleic Acids Research June 15, 2000;28(12):e63.

[32] Notomi T, Mori Y, Tomita N, Kanda H. Loop-mediated isothermal amplification (LAMP): principle, features, and future prospects. Journal of Microbiology January 1, 2015;53(1):1−5.

[33] Matovu E, Kuepfer I, Boobo A, Kibona S, Burri C. Comparative detection of trypanosomal DNA by loop-mediated isothermal amplification and PCR from flinders technology associates cards spotted with patient blood. Journal of Clinical Microbiology June 1, 2010;48(6):2087−90.

[34] Kuboki N, Inoue N, Sakurai T, Cello FD, Grab DJ, Suzuki H, et al. Loop-mediated isothermal amplification for detection of African trypanosomes. Journal of Clinical Microbiology December 1, 2003;41(12):5517−24.

[35] Thekisoe OMM, Kuboki N, Nambota A, Fujisaki K, Sugimoto C, Igarashi I, et al. Species-specific loop-mediated isothermal amplification (LAMP) for diagnosis of trypanosomosis. Acta Tropica June 2007;102(3):182−9.

[36] Desquesnes M, McLaughlin G, Zoungrana A, Dávila AMR. Detection and identification of Trypanosoma of African livestock through a single PCR based on internal transcribed spacer 1 of rDNA. International Journal for Parasitology May 1, 2001; 31(5−6):610−4.

[37] Njiru ZK, Constantine CC, Guya S, Crowther J, Kiragu JM, Thompson RCA, et al. The use of ITS1 rDNA PCR in detecting pathogenic African trypanosomes. Parasitology Research February 1, 2005;95(3):186−92.

[38] Mitashi P, Hasker E, Ngoyi DM, Pyana PP, Lejon V, Veken WV der, et al. Diagnostic accuracy of loopamp *Trypanosoma brucei* detection kit for diagnosis of human African trypanosomiasis in clinical samples. PLoS Neglected Tropical Diseases October 17, 2013;7(10):e2504.

[39] Thekisoe OMM, Bazie RSB, Coronel-Servian AM, Sugimoto C, Kawazu S-I, Inoue N. Stability of Loop-Mediated Isothermal Amplification (LAMP) reagents and its amplification efficiency on crude trypanosome DNA templates. Journal of Veterinary Medical Science April 2009;71(4):471−5.

[40] Batchelor NA, Atkinson PM, Gething PW, Picozzi K, Fèvre EM, Kakembo ASL, et al. Spatial predictions of Rhodesian Human African Trypanosomiasis (sleeping sickness) prevalence in Kaberamaido and Dokolo, two newly affected districts of Uganda. PLoS Neglected Tropical Diseases December 15, 2009;3(12):e563.

[41] Barrett MP. The elimination of human African trypanosomiasis is in sight: report from the third WHO stakeholders meeting on elimination of gambiense human African trypanosomiasis. PLoS Neglected Tropical Diseases December 6, 2018;12(12): e0006925.

[42] Samia Y, Kealey A, Smith? RJ. A mathematical model of a theoretical sleeping sickness vaccine. Mathematical Population Studies April 2, 2016;23(2):95−122.

[43] Berriman M, Hall N, Sheader K, Bringaud F, Tiwari B, Isobe T, et al. The architecture of variant surface glycoprotein gene expression sites in *Trypanosoma brucei*. Molecular and Biochemical Parasitology July 1, 2002;122(2):131−40.

[44] Pinger J, Chowdhury S, Papavasiliou FN. Variant surface glycoprotein density defines an immune evasion threshold for African trypanosomes undergoing antigenic variation. Nature Communications October 10, 2017;8(1):1−9.

[45] Nok AJ. Arsenicals (melarsoprol), pentamidine and suramin in the treatment of human African trypanosomiasis. Parasitology Research January 31, 2003;90(1):71−9.

[46] Priotto G, Kasparian S, Mutombo W, Ngouama D, Ghorashian S, Arnold U, et al. Nifurtimox-eflornithine combination therapy for second-stage African *Trypanosoma brucei gambiense* trypanosomiasis: a multicentre, randomised, phase III, noninferiority trial. The Lancet July 4, 2009;374(9683):56−64.

[47] Anene BM, Onah DN, Nawa Y. Drug resistance in pathogenic African trypanosomes: what hopes for the future? Veterinary Parasitology March 20, 2001;96(2):83−100.

[48] Matovu E, Seebeck T, Enyaru JCK, Kaminsky R. Drug resistance in *Trypanosoma brucei* spp., the causative agents of sleeping sickness in man and nagana in cattle. Microbes and Infection July 2001;3(9):763−70.

[49] Gehrig S, Efferth T. Development of drug resistance in *Trypanosoma brucei rhodesiense* and *Trypanosoma brucei gambiense*. Treatment of human African trypanosomiasis with natural products (Review). International Journal of Molecular Medicine October 1, 2008;22(4):411−9.

[50] Stewart ML, Burchmore RJS, Clucas C, Hertz-Fowler C, Brooks K, Tait A, et al. Multiple genetic mechanisms lead to loss of functional TbAT1 expression in drug-resistant trypanosomes. Eukaryotic Cell February 1, 2010;9(2):336−43.

[51] Mulandane FC, Fafetine J, Abbeele JVD, Clausen P-H, Hoppenheit A, Cecchi G, et al. Resistance to trypanocidal drugs in cattle populations of Zambezia Province, Mozambique. Parasitology Research February 1, 2018;117(2):429−36.

[52] de Koning HP, Jarvis SM. Uptake of pentamidine in *Trypanosoma brucei brucei* is mediated by the P2 adenosine transporter and at least one novel, unrelated transporter. Acta Tropica December 21, 2001;80(3):245−50.

[53] Denise H, Barrett MP. Uptake and mode of action of drugs used against sleeping sickness. Biochemical Pharmacology January 1, 2001;61(1):1—5.

[54] Bray PG, Barrett MP, Ward SA, de Koning HP. Pentamidine uptake and resistance in pathogenic protozoa: past, present and future. Trends in Parasitology May 2003;19(5): 232—9.

[55] Kroubi M, Karembe H, Betbeder D. Drug delivery systems in the treatment of African trypanosomiasis infections. Expert Opinion on Drug Delivery June 1, 2011;8(6):735—47.

[56] Graf FE, Ludin P, Wenzler T, Kaiser M, Brun R, Pyana PP, et al. Aquaporin 2 mutations in *Trypanosoma brucei gambiense* field isolates correlate with decreased susceptibility to pentamidine and melarsoprol. PLoS Neglected Tropical Diseases October 10, 2013;7(10):e2475.

[57] Jansson-Löfmark R, Na-Bangchang K, Björkman S, Doua F, Ashton M. Enantiospecific reassessment of the pharmacokinetics and pharmacodynamics of oral eflornithine against late-stage *Trypanosoma brucei gambiense* sleeping sickness. Antimicrobial Agents and Chemotherapy February 1, 2015;59(2):1299—307.

[58] Alirol E, Schrumpf D, Amici Heradi J, Riedel A, de Patoul C, Quere M, et al. Nifurtimox-eflornithine combination therapy for second-stage gambiense human African trypanosomiasis: Médecins Sans Frontières experience in the Democratic Republic of the Congo. Clinical Infectious Diseases January 15, 2013;56(2):195—203.

[59] Babokhov P, Sanyaolu AO, Oyibo WA, Fagbenro-Beyioku AF, Iriemenam NC. A current analysis of chemotherapy strategies for the treatment of human African trypanosomiasis. Pathogens and Global Health July 1, 2013;107(5):242—52.

[60] Fyfe J, Picozzi K, Waiswa C, Bardosh KL, Welburn SC. Impact of mass chemotherapy in domestic livestock for control of zoonotic *T. b. rhodesiense* human African trypanosomiasis in Eastern Uganda. Acta Tropica January 1, 2017;165:216—29.

[61] Rowland M, Durrani N, Kenward M, Mohammed N, Urahman H, Hewitt S. Control of malaria in Pakistan by applying deltamethrin insecticide to cattle: a community-randomised trial. The Lancet June 9, 2001;357(9271):1837—41.

[62] Torr SJ, Prior A, Wilson PJ, Schofield S. Is there safety in numbers? The effect of cattle herding on biting risk from tsetse flies. Medical and Veterinary Entomology December 1, 2007;21(4):301—11.

[63] Bardosh K, Waiswa C, Welburn SC. Conflict of interest: use of pyrethroids and amidines against tsetse and ticks in zoonotic sleeping sickness endemic areas of Uganda. Parasites and Vectors July 10, 2013;6:204.

[64] Tang W, Wang D, Wang J, Wu Z, Li L, Huang M, et al. Pyrethroid pesticide residues in the global environment: an overview. Chemosphere January 1, 2018;191:990—1007.

[65] Okello-Onen J, Heinonen R, Ssekitto CMB, Mwayi WT, Kakaire D, Kabarema M. Control of tsetse flies in Uganda by dipping cattle in deltamethrin. Tropical Animal Health and Production March 1, 1994;26(1):21—7.

[66] Rogers DJ, Hendrickx G, Slingenbergh JHW. Tsetse flies and their control. Revue Scientifique et Technique Office International des Epizooties 1994;13(4):1075—124.

[67] Rowlands GJ, Coulibaly L, Hecker PA, d'Ieteren GDM, Leak SGA, Authié E. Effect of tsetse control on trypanosome prevalence in livestock: problems of experimental design and statistical interpretation—a case study in northern Côte d'Ivoire. Veterinary Parasitology June 1, 1996;63(3):199—214.

[68] Vale GA, Torr SJ. User-friendly models of the costs and efficacy of tsetse control: application to sterilizing and insecticidal techniques. Medical and Veterinary Entomology 2005;19(3):293—305.

[69] Bekele J, Asmare K, Abebe G, Ayelet G, Gelaye E. Evaluation of Deltamethrin applications in the control of tsetse and trypanosomosis in the southern rift valley areas of Ethiopia. Veterinary Parasitology March 25, 2010;168(3):177—84.

[70] Bauer B, Baumann MPO. Laboratory evaluation of efficacy and persistence of a 1 % w/w fipronil pour-on formulation (Topline®) against *Glossina palpalis gambiensis*, Diptera: Glossinidae. Parasitology Research August 1, 2015;114(8):2919—23.

[71] Vale GA, Hargrove JW, Chamisa A, Grant IF, Torr SJ. Pyrethroid treatment of cattle for tsetse control: reducing its impact on dung fauna. PLoS Neglected Tropical Diseases March 2015;9(3). Available from: https://www.ncbi.nlm.nih.gov/pmc/articles/PMC4349886/.

[72] Sawadogo B, Rayaisse JB, Adakal H, Kabre AT, Bauer B. Fipronil 1% pour-on: further studies of its effects against lab-reared *Glossina palpalis gambiensis*. Parasitology Research 2017;116(11):2927—32.

[73] Abd-Alla AMM, Bergoin M, Parker AG, Maniania NK, Vlak JM, Bourtzis K, et al. Improving Sterile Insect Technique (SIT) for tsetse flies through research on their symbionts and pathogens. Journal of Invertebrate Pathology March 1, 2013;112:S2—10.

[74] Maniania NK, Ekesi S. The use of entomopathogenic fungi in the control of tsetse flies. Journal of Invertebrate Pathology March 1, 2013;112:S83—8.

[75] Hamidou Soumana I, Simo G, Njiokou F, Tchicaya B, Abd-Alla AMM, Cuny G, et al. The bacterial flora of tsetse fly midgut and its effect on trypanosome transmission. Journal of Invertebrate Pathology March 1, 2013;112:S89—93.

[76] Weiss BL, Maltz MA, Vigneron A, Wu Y, Walter KS, O'Neill MB, et al. Colonization of the tsetse fly midgut with commensal *Kosakonia cowanii* Zambiae inhibits trypanosome infection establishment. PLoS Pathogens February 28, 2019;15(2):e1007470.

[77] Carter NS, Fairlamb AH. Arsenical-resistant trypanosomes lack an unusual adenosine transporter. Nature January 14, 1993;361(6408):173—6.

[78] Carter NS, Barrett MP, de Koning HP. A drug resistance determinant in *Trypanosoma brucei*. Trends in Microbiology December 1, 1999;7(12):469—71.

[79] Van Meirvenne N, Magnus E, Büscher P. Evaluation of variant specific trypanolysis tests for serodiagnosis of human infections with *Trypanosoma brucei gambiense*. Acta Tropica December 1, 1995;60(3):189—99.

[80] Schwede A, Jones N, Engstler M, Carrington M. The VSG C-terminal domain is inaccessible to antibodies on live trypanosomes. Molecular and Biochemical Parasitology February 2011;175(2):201—4.

[81] Berriman M, Ghedin E, Hertz-Fowler C, Blandin G, Renauld H, Bartholomeu DC, et al. The genome of the African trypanosome *Trypanosoma brucei*. Science July 15, 2005;309(5733):416—22.

[82] Marcello L, Barry JD. Analysis of the VSG gene silent archive in *Trypanosoma brucei* reveals that mosaic gene expression is prominent in antigenic variation and is favored by archive substructure. Genome Research September 1, 2007;17(9):1344—52.

[83] Woods A, Sherwin T, Sasse R, MacRae TH, Baines AJ, Gull K. Definition of individual components within the cytoskeleton of *Trypanosoma brucei* by a library of monoclonal antibodies. Journal of Cell Science July 1, 1989;93(3):491—500.

[84] Kohl L, Gull K. Molecular architecture of the trypanosome cytoskeleton. Molecular and Biochemical Parasitology May 15, 1998;93(1):1—9.

[85] Lubega GW, Byarugaba-Karuhizel D, Ocholal DO, Prichard RK. Targeting tubulin for vaccine development: immunisation with tubulin from *Trypanosoma brucei* protects mice from infection. South African Journal of Science June 1, 1998;94(6):284—5.

[86] Lubega GW, Byarugaba DK, Prichard RK. Immunization with a tubulin-rich preparation from *Trypanosoma brucei* confers broad protection against African trypanosomosis. Experimental Parasitology September 2002;102(1):9–22.

[87] Rasooly R, Balaban N. Trypanosome microtubule-associated protein p15 as a vaccine for the prevention of African sleeping sickness. Vaccine February 25, 2004;22(8):1007–15.

[88] Li S-Q, Fung M-C, Reid Sa, Inoue N, Lun Z-R. Immunization with recombinant beta-tubulin from *Trypanosoma evansi* induced protection against *T. evansi, T. equiperdum* and *T. b. brucei* infection in mice. Parasite Immunology 2007;29(4):191–9.

[89] Li S-Q, Yang W-B, Lun Z-R, Ma L-J, Xi S-M, Chen Q-L, et al. Immunization with recombinant actin from *Trypanosoma evansi* induces protective immunity against *T. evansi, T. equiperdum* and *T. b. brucei* infection. Parasitology Research 2009;104(2):429–35.

[90] Hoet S, Opperdoes F, Brun R, Quetin-Leclercq J. Natural products active against African trypanosomes: a step towards new drugs. Natural Product Reports May 25, 2004;21(3):353–64.

[91] Hannaert V. Sleeping sickness pathogen (*Trypanosoma brucei*) and natural products: therapeutic targets and screening systems. Planta Medica April 2011;77(6):586–97.

[92] Yabu Y, Nose M, Koide T, Ohta N, Ogihara Y. Antitrypanosomal effects of traditional Chinese herbal medicines on bloodstream forms of *Trypanosoma brucei rhodesiense* in vitro. Southeast Asian Journal of Tropical Medicine and Public Health September 1998;29(3):599–604.

[93] Bawm S, Tiwananthagorn S, Lin KS, Hirota J, Irie T, Htun LL, et al. Evaluation of Myanmar medicinal plant extracts for antitrypanosomal and cytotoxic activities. Journal of Veterinary Medical Science 2010;72(4):525–8.

[94] Nibret E, Wink M. Volatile components of four Ethiopian *Artemisia* species extracts and their *in vitro* antitrypanosomal and cytotoxic activities. Phytomedicine April 1, 2010;17(5):369–74.

[95] Cheraghipour K, Marzban A, Ezatpour B, Khanizadeh S, Koshki J. Antiparasitic properties of curcumin: a review. AIMS Agriculture Food December 21, 2018;3(4):561–78.

[96] Luna EC, Luna IS, Scotti L, Monteiro AFM, Scotti MT, de Moura RO, et al. Active essential oils and their components in use against neglected diseases and arboviruses. Oxidative Medicine and Cellular Longevity 2019;2019:6587150.

[97] Talakal TS, Dwivedi SK, Sharma SR. In vitro and in vivo antitrypanosomal activity of *Xanthium strumarium* leaves. Journal of Ethnopharmacology December 15, 1995;49(3):141–5.

[98] Freiburghaus F, Kaminsky R, Nkunya MHH, Brun R. Evaluation of African medicinal plants for their in vitro trypanocidal activity. Journal of Ethnopharmacology December 1, 1996;55(1):1–11.

[99] Weniger B, Vonthron-Sénécheau C, Kaiser M, Brun R, Anton R. Comparative antiplasmodial, leishmanicidal and antitrypanosomal activities of several biflavonoids. Phytomedicine February 13, 2006;13(3):176–80.

[100] Mann A, Ogbadoyi EO. Evaluation of medicinal plants from Nupeland for their in vivo antitrypanosomal activity. American Journal of Biochemistry August 31, 2012;2(1):1–6.

[101] Polanco-Hernández G, Escalante-Erosa F, García-Sosa K, Acosta-Viana K, Chan-Bacab M, Sagua-Franco H, et al. In vitro and in vivo trypanocidal activity of native plants from the Yucatan Peninsula. Parasitology Research 2012;110(1):31–5.

[102] Changtam C, de Koning HP, Ibrahim H, Sajid MS, Gould MK, Suksamrarn A. Curcuminoid analogs with potent activity against *Trypanosoma* and *Leishmania* species. European Journal of Medicinal Chemistry March 2010;45(3):941–56.

[103] Oliveira IA, Freire-de-Lima L, Penha LL, Dias WB, Todeschini AR. *Trypanosoma cruzi trans*-sialidase: structural features and biological implications. In: Santos ALS, Branquinha MH, d'Avila-Levy CM, Kneipp LF, Sodré CL, editors. Proteins and proteomics of *Leishmania* and *Trypanosoma*. Subcellular biochemistry, vol. 74. Dordrecht, Netherlands: Springer; 2014. p. 181–201. https://doi.org/10.1007/978-94-007-7305-9_8. Available from:.

[104] Kashif M, Moreno-Herrera A, Lara-Ramirez EE, Ramírez-Moreno E, Bocanegra-Garcia V, Ashfaq M, et al. Recent developments in trans-sialidase inhibitors of *Trypanosoma cruzi*. Journal of Drug Targeting July 3, 2017;25(6):485–98.

[105] Nardy AFFR, Freire-de-Lima CG, Pérez AR, Morrot A. Role of *Trypanosoma cruzi trans*-sialidase on the escape from host immune surveillance. Frontiers in Microbiology 2016;7. Available from: https://www.frontiersin.org/articles/10.3389/fmicb.2016.00348/full.

[106] Arias DG, Herrera FE, Garay AS, Rodrigues D, Forastieri PS, Luna LE, et al. Rational design of nitrofuran derivatives: synthesis and valuation as inhibitors of *Trypanosoma cruzi* trypanothione reductase. European Journal of Medicinal Chemistry January 5, 2017;125:1088–97.

[107] Mendonça AAS, Coelho CM, Veloso MP, Caldas IS, Gonçalves RV, Teixeira AL, et al. Relevance of trypanothione reductase inhibitors on *Trypanosoma cruzi* infection: a systematic review, meta-analysis, and in silico integrated approach. Oxidative Medicine and Cellular Longevity 2018;2018:8676578.

[108] Daunes S, Yardley V, Croft SL, D'Silva C. Antiprotozoal glutathione derivatives with flagellar membrane binding activity against *T. brucei rhodesiense*. Bioorganic and Medicinal Chemistry February 15, 2017;25(4):1329–40.

[109] Galarreta BC, Sifuentes R, Carrillo AK, Sanchez L, Amado M del RI, Maruenda H. The use of natural product scaffolds as leads in the search for trypanothione reductase inhibitors. Bioorganic and Medicinal Chemistry July 15, 2008;16(14):6689–95.

[110] Leroux AE, Krauth-Siegel RL. Thiol redox biology of trypanosomatids and potential targets for chemotherapy. Molecular and Biochemical Parasitology March 2016;206(1–2):67–74.

Index

Printed in the United States
By Bookmasters